Springer Series in Materials Science

Volume 193

For further volumes:
http://www.springer.com/series/856

The Springer Series in Materials Science covers the complete spectrum of materials physics, including fundamental principles, physical properties, materials theory and design. Recognizing the increasing importance of materials science in future device technologies, the book titles in this series reflect the state-of-the-art in understanding and controlling the structure and properties of all important classes of materials.

Alexander Ziegler · Heinz Graafsma
Xiao Feng Zhang · Joost W. M. Frenken
Editors

In-situ Materials Characterization

Across Spatial and Temporal Scales

 Springer

Editors

Alexander Ziegler
Microscopy and Microanalysis Unit
The University of the Witwatersrand
Johannesburg
South Africa

Xiao Feng Zhang
Hitachi High Technologies America, Inc.
Nanotechnology Systems Division
Pleasanton, CA
USA

Heinz Graafsma
Photon-Science Detector Group
Deutches Elektronen-Synchrotron, DESY
Hamburg
Germany

Joost W. M. Frenken
Kamerlingh Onnes Laboratory
Leiden University
Leiden
The Netherlands

ISSN 0933-033X   ISSN 2196-2812   (electronic)
ISBN 978-3-642-45151-5   ISBN 978-3-642-45152-2   (eBook)
DOI 10.1007/978-3-642-45152-2
Springer Heidelberg New York Dordrecht London

Library of Congress Control Number: 2014934837

Printed on acid-free paper

Springer is part of Springer Science+Business Media (www.springer.com)

Preface

Congratulations on selecting this book.

Now, what are you expecting from this book? The title suggests a scientific content and this may have caught your attention. Perhaps the first words *"In-situ"* have made you look closer at this book. But, what does *"In-situ"* really mean? *"In-situ"*—a Latin phrase—translates to *"in position"* or *"in place."* Hence, *"In-situ Characterization,"* therefore means nothing but *"observing while exactly in place where things occur."* These "things" can be reactions, transformations, alterations, and/or changes to a state of equilibrium of matter, or a sample of it. Very often such changes in equilibrium are unstable and only transient in nature. Consequently, they cannot be isolated, or observed before and after such a reaction or transformation. A meticulous examination of these transient stages of dynamic processes is therefore the aim of numerous scientific investigations.

There are a number of techniques—characterization techniques to be precise—that in the past 100 years have set off a tremendous growth and accumulation of scientific knowledge and insight into basically every aspect of applied and theoretical sciences. Let us consider this the static and motionless type of characterization. In 1895 Wilhelm C. Röntgen discovered X-rays. In 1932 Ernst A. F. Ruska invented the electron microscope. In 1946/1947 the synchrotron radiation was discovered and sufficiently high neutron fluxes for neutron scattering became available around the same time with the advent of nuclear reactors. In 1960 the first functional laser was in operation at Bell Labs, and in 1981 the first scanning tunneling microscope was invented at IBM in Zurich. The growth, development, and diversification of any of these and other characterization techniques has led to many groundbreaking discoveries and phenomenal structural and analytical results, often revealing fascinating aspects of matter, nature, and life. Indirect, but more importantly the direct visualization of individual atoms and their structural and electronic configuration in two and three dimensions in space, revealing constitutive laws of a predominant component of matter, is just one phenomenal example of the power of these (motionless) characterization techniques. However, seeing these essential "building blocks" of nature and their constructs at the nanometer and micrometer scale react to external forces, caused for example by changes in their thermodynamic state, invariably raises our level of understanding of the dynamic aspects of nature—we are adding the fourth dimension: time.

Time-resolved studies *per se* are not new. In 1878 Eadweard J. Muybridge's sequence of high-speed photographic recordings revealed for the first time how a galloping race horse has all four hoofs in the air at once. In 1886 Peter Slacher photographed the first supersonic flying bullet. Today, the dynamics of extremely fast processes at the macroscopic scale can be examined by photographically recording approximately one-trillion frames per second—Femto-photography.

Thus, coming back to the "In-situ" aspect, in-situ characterization is therefore nothing but enabling the observation of dynamic processes—*"in place"*—at the micrometer, nanometer, and even atomic scale, by performing time-resolved studies ranging from many hours down to femtoseconds—in some cases even attoseconds. The latter ones pertain to the field of ultrafast science—an emerging field of research with many experiments-in-waiting to be addressed or revisited at a different, much faster timescale. A number of very recent technical and instrumentation developments can be expected to enable more rapid advances in the next decade, while other aspects of ultrafast science are representing severe obstacles that may require some more time until a solution is found. Similar scenarios apply to the "slower" time-resolved characterization methods that are also described in this book, and where the primary focus is not the time aspect. Here the challenges are rather sample and sample environment related, e.g., gaseous or liquid sample, temperature, pressure or electrical conductivity, etc.

This book is certainly not fully comprehensive, and it is meant to give the reader an overview of some of the time-resolved techniques and methods that exist, and what is currently the state of the art. Examples are presented that give the reader insight into what type of experiments are possible with different techniques, what the results look like, and how they are interpreted. Technical and methodological capabilities, but more importantly the controlling and the limiting parameters of each technique are—as usual—the critical elements that decide whether or not a particular experiment can be performed and what type of information and result can be expected.

The idea for this book started at a Materials Research Society meeting in Boston, where a strong interest in the latest developments and experimental capabilities for in-situ and time-resolved studies was continuously demonstrated. The individual chapters deal with time-resolved studies using X-ray absorption spectroscopy, Synchrotron and Free Electron Laser X-ray diffraction, Transmission Electron Microscopy, Neutron Spectroscopy and Scanning Tunneling Microscopy. An additional final chapter describes the detector types for X-rays and electrons that have to follow progress on the source instrumentation and experimental setup side. No suitable detector—no spectrum or image.

Johannesburg	Alexander Ziegler
Hamburg	Heinz Graafsma
Pleasanton	Xiao Feng Zhang
Leiden	Joost W. M. Frenken

Contents

Contributors

Majed Chergui Laboratoire de Spectroscopie Ultrarapide ISIC, Faculté des Sciences de Base, Ecole Polytechnique Fédérale de Lausanne, Station 6, Lausanne, Switzerland, e-mail: majed.chergui@epfl.ch

Götz Eckold Institut für Physikalische Chemie, Georg-August-Universität Göttingen, Tammannstrasse 6, 37077 Göttingen, Germany, e-mail: geckold@gwdg.de

Joost W. M. Frenken Kamerlingh Onnes Laboratory, Leiden University, Niels Bohrweg 2, 2333 CA Leiden, The Netherlands, e-mail: Frenken@Physics.LeidenUniv.nl

Heinz Graafsma Photon-Science Detector Group, Deutches Elektronen-Synchrotron, DESY, Notkestr. 85, D-22607 Hamburg, Germany, e-mail: heinz.graafsma@desy.de; STC Research Center, University of Mid-Sweden, 85107 Sundsvall, Sweden, e-mail: heinz.graafsma@desy.de

Helmut Schober Science division, Institut Laue-Langevin, 6, rue Jules Horowitz, BP 156, 38042 Grenoble Cedex 9, France, e-mail: schober@ill.eu

Vedran Vonk Institute for Molecules and Materials, Radboud University Nijmegen, Nijmegen, The Netherlands, e-mail: vedran.vonk@desy.de; Photon-Science Nanolaboratory, Deutches Elektronen-Synchrotron, DESY, Hamburg, Germany, e-mail: vedran.vonk@desy.de

Xiao Feng Zhang Nanotechnology Systems Division, Hitachi High Technologies America, Inc., 5960 Inglewood Drive, Suite 200, Pleasanton, CA 94588, USA, e-mail: xiao.zhang@hitachi-hta.com; Hitachi High Technologies Corporation, Tokyo 105-8717, Japan, e-mail: xiao.zhang@hitachi-hta.com

Alexander Ziegler Microscopy and Microanalysis Unit, The University of the Witwatersrand, 1 Jan Smuts Ave., Johannesburg 2000, South Africa, e-mail: alexander.ziegler@wits.ac.za

Chapter 1
In-situ Characterization of Molecular Processes in Liquids by Ultrafast X-ray Absorption Spectroscopy

Majed Chergui

Abstract The need to visualize molecular and electronic structure in the course of a chemical reaction, a phase transformation a biological function has been the dream of scientists for decades. The development of time-resolved X-ray and electron based methods is making this true. X-ray absorption spectroscopy is ideal for the study of structural dynamics in liquids, because it can be implemented in amorphous media and it is chemically selective. Using X-ray absorption near edge structure (XANES) and extended X-ray absorption fine structure (EXAFS) in laser pump/X-ray probe experiments allows the retrieval of the local geometric structure of the system under study, but also the underlying photoinduced electronic structure changes that drive the structural dynamics. We review the recent development in picosecond and femtosecond X-ray absorption spectroscopy applied to molecular systems in solution: examples on ultrafast photoinduced processes such as intramolecular electron transfer, high-to-low spin change, bond formation and water dynamics are presented.

1.1 Introduction

The advent of femtosecond laser spectroscopy made it possible to observe atomic motions on a time scale that is shorter than a single vibrational period in molecular systems, [1, 2] giving birth to the field of Femtochemistry, which led to the Nobel Prize for Chemistry to Ahmed. H. Zewail in 1999. With femtochemistry, it became possible to follow in real-time molecular and biochemical reactions, phase transitions or coherent oscillations in solids, opening immense insight in our understanding of light-induced phenomena in matter. The impact was huge in

M. Chergui (✉)
Laboratoire de Spectroscopie Ultrarapide ISIC, Faculté des Sciences de Base,
Ecole Polytechnique Fédérale de Lausanne, Station 6, Lausanne, Switzerland
e-mail: majed.chergui@epfl.ch

A. Ziegler et al. (eds.), *In-situ Materials Characterization*,
Springer Series in Materials Science 193, DOI: 10.1007/978-3-642-45152-2_1,
© Springer-Verlag Berlin Heidelberg 2014

condensed matter physics, in chemistry and in biology, as witnessed to this day by the impressive activity in ultrafast spectroscopy. In femtosecond spectroscopy, a first (pump) laser pulse excites the system at time zero, while a second (probe) pulse probes its evolution as a function of time delay with respect to the pump pulse. However, optical domain spectroscopy does not deliver structure. Thus already in the early days of Femtochemistry, efforts were deployed aimed at implementing the traditional structural tools of X-ray and electron diffraction and of X-ray absorption spectroscopy in time-domain experiments, towards the time resolution of picoseconds to femtoseconds.

The great advantage of electrons is that they scatter on matter 5–6 orders of magnitude more efficiently than X-rays [3]. The techniques of ultrafast electron diffraction, crystallography and microscopy have been pioneered by Ahmed H. Zewail and co-workers, [4–11] who showed the potential and capabilities of these techniques for studying the dynamics of gas phase molecules, surfaces, interfaces, biological systems and in materials science. In particular, the tools of ultrafast electron microscopy have heralded a real revolution by allowing real-time and real-space imaging of photoinduced effects. In addition, novel methods keep being added to the tool-kit of electron-based techniques, such as ultrafast EELS, [12, 13] which now allows the visualization of the electronic structure changes together with the geometric changes and therefore relate the two or the Photoinduced near-field electron microscopy (PINEM) approach, [14] which allows visualization of the electric field around photoexcited species. New developments are also possible using relativistic ultrafast electrons to probe matter [15]. In summary, electron-based ultrafast research is a new area of research with ramifications into very diverse fields thanks to a broad range of methods [16]. Last but not least, electron techniques are all lab-based, which offers a great flexibility and versatility.

The high scattering cross-section of electrons versus X-rays, also implies that they are not very penetrating (except for relativistic ones) compared to the latter, so that they are not ideal to probe bulk dynamics, for which ultrashort X-ray pulses seem more appropriate. There are two main classes of sources of ultrashort X-ray pulses: lab-based ones and large-scale installations, such as synchrotrons and very recently, X-ray free electron lasers (X-FEL). The lab-based sources mainly consist of plasmas, which are generated when an intense femtosecond laser pulse hits a metal target, delivering X-ray pulses of 100's fs. However, the ensuing X-ray emission is isotropic, polychromatic, low flux, and the source is unstable on a pulse-to-pulse basis. Synchrotron sources provide fairly high fluxes of very stable, tunable and collimated X-rays, but the pulse width is in the 50–100 ps range. They have been used at this time resolution for molecular and protein crystals [17–23] as well as for scattering studies of molecules in solution [24–30]. Recently, with the advent of the slicing scheme, [31–34] femtosecond pulses of monochromatic X-ray photons are also available at synchrotrons, that have been used for both X-ray diffraction studies [35] and X-ray absorption spectroscopy ones [36–39].

Ultrafast X-ray diffraction has been used to study coherent lattice dynamics, strain propagation, melting phenomena and phase transitions in bulk materials and in nanostructures [35, 40–47]. As mentioned above, protein and molecular crystals

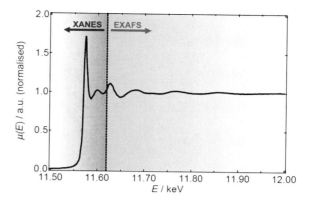

Fig. 1.1 X-ray absorption spectrum at the L_3 edge of a diplatinum complex (Pt₂POP) in solution [206] illustrating the low-energy XANES *region* up to ~50 eV above the ionization potential (IP) and the high-energy EXAFS *region* >50 eV above the IP. The first peak at the edge is due to a bound–bound 2p-5d transition, [207] and the IP lies above it

have been investigated with 100 ps to ns time resolution, while the first studies at higher temporal resolution are starting to appear on such systems [48–50]. X-ray diffraction of liquids initially proposed by Wilson and co-workers [51–53] has so far been limited to the 100 ps resolution [29, 54] and is challenging due to heating, lack of long-range order, the presence of several photoproducts and the high background of light scattered by the solvent and the unexcited species. The studies carried out at synchrotrons are however crucial as they prepare the ground for the implementation of similar experiments with femtosecond resolution at the new and intense sources of X-rays, such as the X-ray free electron lasers (XFELs).

Chemistry and biology mainly occur in the condensed phases and in particular, liquids. It is thus important to probe the chemical and biochemical dynamics in this environment. Furthermore, molecular (geometric) structure changes are triggered and/or accompanied by electronic structure changes, either under excitation by light or in the presence of a reactant. Ideally, one would like not only to probe the *molecular structure* changes in real-time, but also identify the *electronic structure* changes underlying them. In recent years, time-resolved X-ray absorption spectroscopy (XAS) has emerged as an ideal tool for the study of structural dynamics in liquids [10, 30, 36, 55–61]. The present contribution is about these developments and new ones to come. We purposely limited ourselves to studies carried out at fs to sub-ns resolution.

1.2 X-ray Absorption Spectroscopy

X-ray absorption spectra are characterized by absorption edges, which are element specific and reflect the excitation of core electrons to the ionization threshold [62–64]. At a given edge, the spectrum consists of bound–bound transitions below the ionization potential (IP) of the inner-shell electron, followed by the absorption edge jump itself (Fig. 1.1). Right above the absorption edge one observes a complicated modulation of the absorption cross section due to multiple scattering of the ionized photoelectron

Fig. 1.2 Pictorial view of
the single- and multiple-
scattering pathways of an
outgoing photoelectron wave
(*blue*) off 1st-shell (*red*) and
2nd-shell (*green*) neighboring
atoms. The strength of the
scattered wave is reflected in
the line thickness

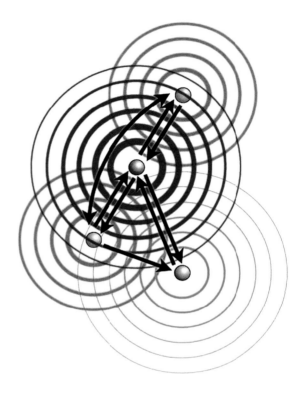

by its nearest neighbors in the molecule. The spectral range around the edge is called
the XANES region (X-ray absorption near-edge structure), and it contains rich information about both the electronic (below the IP) and the geometric structure (above
the IP) including bond angles and distances (Fig. 1.1).

From about 50 eV above IP to higher energies mainly single scattering events
dominate, which result in a weak oscillatory modulation of the absorption cross
section. This region is called the EXAFS region (Extended X-ray Absorption Fine
Structure). It delivers precise information about the internuclear distance of the
nearest neighbors from the absorbing central atom. Oscillations therefore do not
exist in the case of the isolated atom. Thus a single spectrum contains simultaneously information about the valence electronic structure and about the nuclear
arrangement of the atoms in the molecular system (Fig. 1.2).

1.2.1 Extended X-ray Absorption Fine Structure (EXAFS)

In the energy regime of EXAFS, the photoelectron states approximate to spherical waves. EXAFS is a final-state quantum-interference effect involving scattering of the outgoing photoelectron from the neighbouring atoms, and the oscillatory

structure is due to the interference between the outgoing photoelectron wave and the wave scattered back at neighboring atoms (Fig. 1.2). At high photoelectron kinetic energies, the scattering of electrons is such that the only significant contributions to the final state wave function in the vicinity of the absorbing atom comes from paths, in which the electron is scattered only once (single scattering events) [65]. The photoelectrons emitted from the excited atom as spherical waves damp out rapidly due to inelastic effects caused by the extended valence orbitals of the nearby-lying atoms. This limits the probed spatial region and, it ensures that multiple-scattering effects beyond simple back-scattering can be ignored. This allows the analysis of the data by a simple Fourier transformation [66]. For this purpose one has to generate an X-ray absorption spectrum, normalized to the absorption edge jump under consideration, which is defined as the normalized oscillatory part of $\mu(E)$ (the X-ray absorption coefficient), i.e. the EXAFS, via

$$\chi(E) = \frac{[\mu(E) - \mu_0(E)]}{\Delta\mu_0(E)} \tag{1.1}$$

with $\mu_0(E)$ being the smoothly varying atomic-like background absorption and $\Delta\mu_0$ is a normalization factor that arises from the net increase in the total atomic background (or simply the absorption edge jump). Substituting E above the edge with the photoelectron wave vector, we can rewrite $\chi(k)$ as

$$\chi(k) = \sum_j S_0^2 N_j \frac{|f_j(k)|}{kR_j^2} \sin\left[2kR_j + 2\delta_e + \Phi\right] \cdot e^{-\frac{2R_j}{\lambda(k)}} \cdot e^{-2\sigma_j^2 k^2} \tag{1.2}$$

which is the standard EXAFS formula. The structural parameters (for which the subscript j refers to the group of N_j atoms with identical properties, e.g., bond distance and chemical species) are:

(a) the interatomic distances R_j,
(b) the coordination number (or number of equivalent scatterers) N_j,
(c) the temperature-dependent *rms* fluctuation in bond length σ_j, which should also include effects due to structural disorder.
(d) $f_j(k) = |f_j(k)|e^{i\phi(k)}$ is the backscattering amplitude, δ_e is central-atom partial wave phase shift of the final state, and $\lambda(k)$ is the energy-dependent photoelectron mean free path (not to be confused with its de Broglie wavelength), and S_0^2 is the overall amplitude factor.

Moreover, although the original EXAFS formula referred only to single-scattering contributions from neighboring shells of atoms, the same formula can be generalized to represent the contribution from N_R equivalent multiple-scattering contributions of path length $2R$ [64].

Equation 1.2 contains all of the key elements that provide a convenient parametrization for fitting the local atomic structure around the absorbing atom to the measured EXAFS data. The dependence of the oscillatory structure on interatomic distance and energy is clearly reflected by the $\sin(2kR)$ term. The decay of the wave due to the mean free path or finite lifetime (including the ultrashort core–hole lifetime) of the photoelectron is captured by the exponential term $e^{-2R/\lambda}$. The strength of the reflected interfering waves depends on the type and number of neighboring atoms and is given by the backscattering amplitude $|f_j(k)|$ and hence is primarily responsible for the intensity of the EXAFS signal. Other factors, namely, the spherical-wave factors $(1/kR^2)$ and mean-free-path terms, appear secondary but are important for a quantitative description of the EXAFS amplitude. The phase factor $\Phi = \arg f(k)$ reflects the quantum–mechanical wavelike nature of the backscattering process. The $\exp(-2\sigma^2 k^2)$, representing the Debye–Waller broadening to a good approximation, is partly due to thermal effects, which cause the atoms to jiggle around their equilibrium atomic positions. These slight movements smear out the otherwise sharp interference pattern of the rapidly varying $\sin(2kR)$ term with increasing k. Effects of structural disorder are similar and they give an additive contribution to σ^2. The Debye–Waller effect seems more pronounced at shorter photoelectron wavelengths and hence it terminates the EXAFS at sufficiently large energy beyond $k \sim 1/\sigma$, which is typically around 10 Å^{-1}. Finally, S_0^2 is a many-body effect due to the relaxation of the system in response to the creation of the core hole. It is usually approximated by a constant. Using 1.2, it can be seen that a Fourier transform of the EXAFS with respect to k corresponds to an effective radial distribution function, with peaks near the first few nearest neighbor distances, provided the phase shifts are correctly dealt with. Other improvements to the theory have been amended, such as multiple scattering and curved-wave effects [64].

For disordered system, Filipponi [67] has derived an equivalent of (1.2) that uses an integral over the radial distribution function (g) rather than a discrete sum:

$$\chi(k) = \sum_i \int_0^\infty 4\pi \rho_i r^2 g_{X-S_i}(r) A_i(k,r) \sin(2kr + \varphi_i(k,r)) dr \qquad (1.3)$$

where X is the absorber (e.g., solute), i stands for a given atom of the solvent species. $A(k,r)$ and $\phi(k,r)$ are the amplitude and phase functions. An average solvation shell structure, described by radial pair distribution functions between the (in this case, atomic) solute and each class of solvent atom, can be extracted from the analysis of the EXAFS signal using this equation. Theoretical EXAFS signals associated with each type of solvent atom of the first shell molecules, are calculated from (1.3) using standard packages. The phase shifts $\phi(k,r)$ are calculated from a snapshots of atomic coordination given by molecular dynamics simulations. The $\chi(k)$ signal is mainly sensitive to a limited distance from the absorber (A) due to the finite photoelectron mean-free path, this effect together with the spectral damping due to monochromator resolution are included in the scattering amplitude $A(k,r)$.

1.2.2 X-ray Absorption Near Edge Spectroscopy (XANES)

In contrast to the EXAFS region, for which accurate quantitative results are generally routine, both the calculation and interpretation of XANES have remained challenging tasks. The complications arise from the low kinetic energy of the photoelectron in this region resulting in large scattering amplitudes (and consequently multiple scattering events, Fig. 1.2) and the necessity to adequately describe many-body effects. In addition, the photoelectron energy is comparable to the fine details of the atomic Coulomb potential which renders the XANES spectrum very sensitive to inaccuracies of the simplified model that is chosen to represent the atomic potentials. With the development of an efficient self-consistent, relativistic and full multiple scattering code (FEFF) by Rehr and coworkers, [64, 68, 69] the theoretical errors in XANES calculations are largely reduced making a (semi-) quantitative interpretation of the spectral features possible. This is achieved in particular by the efficient many-pole representation of the system dielectric function in order to account for intrinsic and extrinsic inelastic losses of the photoelectron [70] and the possibility to use full potentials instead of the simplified muffin-tin (MT) approximation [71, 72].

In general, XAS, in the single electron and dipole approximations, probes the probability of a dipole-mediated transition of a deep core electron $|i\rangle$ of the absorbing atom into some unoccupied state $|f\rangle$ above the Fermi level E_F. The measured X-ray absorption cross section $\mu(E)$ is described by the Fermi golden rule

$$\mu(E) \propto \sum_f^{E_f > F_F} \left| \left\langle f \left| \hat{\varepsilon} \cdot \vec{r} \right| i \right\rangle \right| \delta \left(E - E_f \right) \tag{1.4}$$

where $\hat{\varepsilon} \cdot \vec{r}$ is the dipole operator of the incoming light and the sum is over all energies above the Fermi level. There are two general strategies for solving this equation in order to calculate the XAS spectrum [70, 73]. The first implies accurately expressing the deep core state $|i\rangle$ and final states $|f\rangle$ in terms of wave functions and then explicitly evaluating the sum of (1.4). This is essentially the approach adopted by molecular orbital theories, but it is limited by the accurate description of the final state. This limitation is avoided by the second strategy adopted by multiple scattering theories for which (1.4) is rewritten using a single-particle Green's function operator $G = [E - H]^{-1}$, where H is the effective one-electron Hamiltonian and E is the photoelectron energy. The use of the Green's function formalism as a tool to avoid the explicit representation of final states is the basis of Multiple Scattering theory which originated in the 1970s [74, 75]. One of the approximations is to write the total Hamiltonian of the system as $H = H_0 + V_{int} + \delta V$, where V_{int} is a flat interstitial potential due to the system of ions and electrons in a given system. The ions are the scattering sites of potential δV. The scattering atomic potentials H_0 are approximated with spherically symmetric Coulomb-type potentials embedded within the interstitial potential V_{int}. This approximation is called *muffin-tin* (MT) potential approximation. Within the MT approximation, the electron wavefunction can be expressed by spherical harmonics with an additional radially dependent part, which extends into the V_{int} region. The outer region sets the specific boundary conditions onto the electron wave function acting within the cluster sketched. The Multiple Scattering equations derived from this approach are the basic ingredients of the computer codes used in XAS calculations, which consist of mainly two families: the FEFF and

MXAN codes. Both programs allow computing the XAS spectra within the MT approximation used in Multiple scattering theory and in addition they can be used to fit the experimental spectra given the initial structure of the atomic cluster [68, 74, 76].

Several key developments have increased the accuracy of the calculations, offering the possibility to interpret XANES either in terms of local electronic structure or within a XAFS-like scattering picture [70]: (1) the use of full multiple-scattering (FMS) cluster calculations of both XAS and DOS; (2) the use of self-consistent field (SCF) potentials (both of these developments are based on Green's function calculations); (3) ab initio calculations of many-body intrinsic and extrinsic inelastic losses of the photoelectron. More details are found in the cited papers, in particular [69].

A crucial point in all time-resolved X-ray studies is the extraction of the excited state spectrum from the transient (difference transmission with the above described scheme) spectra, which is a prerequisite for the structural analysis. Naturally, this is strongly sensitive to the precise determination of the photoexcitation yield, as this may induce errors in the structural determination [58]. In the simple case of a mere two-level system, the analysis uses the XAS spectrum of the ground and (extracted) excited state. Several codes exist that fit both the XANES [64, 68, 70, 77–81] and the EXAFS [64, 82] regions. However, quite frequently it is difficult, if not impossible, to extract the excited state fractional population from comparative optical-only studies, because the ground state bleach (GSB) is often overlapped by excited state absorption (ESA) or stimulated emission (SE). In addition, populations extracted from laser-only studies need to be convoluted with the much longer X-ray pulse width, giving rise to additional ambiguities.

An alternative approach was developed that fits the transient EXAFS spectrum in energy space directly [83]. A series of EXAFS spectra are simulated for a collection of possible excited state structures from which the ground state fit spectrum is subtracted to generate transient spectra. These are then compared with the experimental TA spectrum using a least-squares statistical analysis to derive the structural change. This approach reduces the number of required parameters by cancellation in the differences. It can also deliver a unique solution for both the fractional population and the extracted excited state structure, next to quantifying electronic information about possible energy shifts [83]. A similar approach was also proposed for fitting difference XANES spectra [80]. It is however clear that one needs to have a preconceived idea of the input structures, which are based on knowledge of the system and its realistic deformation geometries, as for example in the case of bimetallic complexes (see Sect. 1.4.2).

Concerning the electronic structure information from the bound–bound XANES features, the charge transfer ligand field multiplet theory [84–87] has proven very successful on a wide range of systems, both in molecular and condensed matter physics. The theory takes into account all Coulomb interactions as well as the spin–orbit coupling between the atomic orbitals, and treats the geometric environment of the absorbing atom through a ligand field potential. It also introduces orbital mixing between the central atom and its ligands via charge transfer terms. This tool has now been extended to model excited state XANES spectra [88] but here too, the comparison with experiment is critically dependent on a most accurate determination of the fraction of excited state population [58].

Summarising this section, XANES and EXAFS are useful to extract the following type of information:

a. The bound–bound XANES features are due to transitions from core orbit also to valence ones, and the latter are involved in chemical bonding and/or are transformed under visible-ultraviolet light excitation. Thus one can interrogate the occupancy of valence orbitals, the oxidation state of the system, the occurrence of charge transfer processes, orbital delocalization, bonding and backbonding contributions, etc. [63, 87, 89], and their changes in the course of a chemical, biochemical or physical transformation.
b. Above ionization XANES multiple scattering resonances (also called shape resonances [63]) can deliver information about bond distances and angles, as well as coordination numbers [71, 90–92].
c. The single scattering EXAFS region delivers information about bond distances and coordination numbers of the nearest neighbours around the absorbing atom. The structural analysis of EXAFS is much simpler than that of XANES, which is the reason why EXAFS is more commonly used in static studies of molecular structure, even though the EXAFS modulations are typically less than 10 % of the edge jump.

Furthermore, XAS (XANES and EXAFS) offers the following advantages:

i.   it is atom selective
ii. it can detect species that are spectroscopically silent in the optical domain
iii. it can be implemented in any phase of matter (gas, liquid, solid) and in biological systems
iv. it delivers information about both the electronic and the geometric structure of the system under study
v.   the information about geometric changes is local (i.e. one to three shells of neighbours around the species of interest), but this is not a problem since ultrashort time scales correspond to ultrashort distance changes.

In the past 10 years, time-resolved XAS of liquids systems has matured to a routine method on the picosecond time scale, [10, 55–58, 93] while the femtosecond time resolution has just been achieved, [36, 39] opening exciting perspectives for ultrafast structural dynamics of chemical and biological systems in liquids.

1.3 Methodology

For time-resolved X-ray absorption spectroscopy, the X-ray source should ideally fulfill the following requirements: (1) continuum radiation covering the range from soft to hard X-ray energies; (2) ultrashort pulses; (3) high photon fluxes; (4)

Fig. 1.3 experimental lay
out of an optical pump/X-ray
absorption probe experiment
at a synchrotron

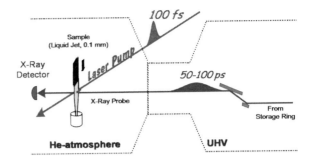

high pulse-to-pulse, as well as long term stability; (5) synchronization to an external light source (e.g., a pump laser). The only X-ray sources that approach these requirements are 3rd generation synchrotron storage rings, albeit at the cost of a lower temporal resolution (tens of ps). However, the sub-ps range has been achieved at synchrotrons thanks to the implementation of the so-called slicing scheme [31, 33, 34, 94] though at a largely reduced photon flux.

There has been several attempts to use plasma sources of X-rays for ultrafast XAS experiments ever since the beginning of the 1990s, [95–104] but these never materialised into a routine methodology as much as the synchrotron-based approach did. This has to do with the experimental difficulties and limitations inherent to plasma sources that have large pulse to pulse fluctuations, low fluxes at a given energy bandwidth as well as limited spectral tunability. Synchrotron based X-ray absorption experiments have had more success and are now routinely used at the SLS (Villigen), APS (Argonne), ALS (Berkeley) and BESSY (Berlin), while more experiments are being implemented. However, plasma-based sources have had (and still have) much success in ultrafast X-ray diffraction studies [42, 43, 45, 105].

The development and methodology of time resolved XAS studies at synchrotrons have been described in refs [10, 55–59, 106–110]. The now commonly used approach was implemented by the Lausanne group. It is based on recording the transient (difference) absorption on a pulse-to-pulse basis. Briefly (Fig. 1.3), monochromatic X-rays are focussed to 50–100 μm diameter onto the sample. X-ray signals are detected with large-area avalanche photodiodes (APD), of which one for transmission (I_1) and one for the incident signal (I_0) scattered off a thin metal foil, and two (or more, if space allows) fluorescence APDs (I_{F1}, I_{F2}) for the X-ray fluorescence from the sample. The sample consists of a free-flowing liquid jet, which is excited by an ultrashort laser pulse. Spatial overlap including a measurement of the spot sizes on the sample between both laser and X-ray beams is set via steering the laser beam onto the X-ray spot on sample, monitored with an imaging CCD camera.

These laser-pump/X-ray probe experiments at synchrotrons exploit a special electron bunch filling pattern (Fig. 1.4) such as that found at the ALS (Berkeley, USA) or the SLS (Villigen, Switzerland), but other filling modes are convenient,

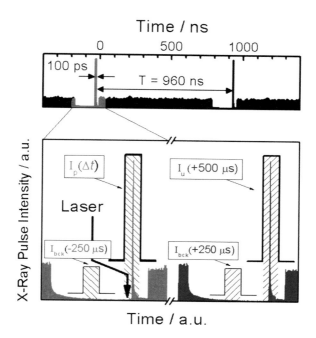

Fig. 1.4 *Top* Typical bunch filling pattern at synchrotrons. Here we show the case of the Swiss *Light Source* Trains of 390 electron bunches with 2 ns spacing are followed by a 180 ns long empty section, in which a densely packed electron bunch (pulse width 100 ps) is placed 30 ns before the start of the bunch train. *Bottom* Data acquisition scheme employed at the microXAS beamline of the Swiss Light Source. A gated integrator measures sequentially the X-ray intensity at the chosen time delay Δt, a background signal ca. 250 μs earlier and later, the X-ray signal of the unexcited sample after 500 μs (each within the indicated shaded time windows). This acquisition scheme is repeated for several thousand times per data point

such as that used at the Advanced Photon Source (APS, Argonne), provided individual pulses can be isolated.

For example, the SLS storage ring consists of 480 so-called buckets, each separated by 2 ns, in which 390 are filled (a filled bucket is also called an electron bunch), while in the subsequent 180 ns dark section a single densely packed (with up to 5 times more charge) hybrid electron bunch is placed. The X-ray pulse emitted from this electron bunch is used to probe the photoinduced changes of the X-ray absorption within the sample. In most experiments so far carried out, amplified laser systems running at 1 kHz were used for photoexcitation, and are synchronized to the round trip frequency of the electron bunch (T = 960 ns at the SLS), thus ensuring a constant time delay between the laser and selected probing X-ray pulses. A gated integrator delivers a sequence of output signals proportional to the input signal integrated over a fixed time window (Fig. 1.4). These signals are recorded with an ADC card, triggered at 4 kHz, to provide a sequence of the X-ray signals when the laser is on and off, but also for recording the electronic background signal with no X-rays present. First, the X-ray signal at time Δt after laser excitation is measured (I_p), then a background signal without X-rays is recorded after ca. 250 μs (I_{bck}(250 μs)), which enables post-correcting baseline drifts in the gated integrator, and finally, after 0.5 ms, the X-ray signal from a fresh sample without laser excitation is recorded (I_u) followed by a baseline correction measurement (I_{bck}(500 μs + 250 μs), not shown in the figure) [93, 110]. The pump-probe signal in transmission is defined as the transient absorption T_{tr} of the photoexcited sample via [93, 110].

$$T_{tr}(E, \Delta t) = \ln\left(\frac{I_{pum}}{I_{unp}}\right) \tag{1.5}$$

with $I_{pum} = I_p(\Delta t) - I_{bck}(\Delta t + 250$ μs) and $I_{unp} = I_u(500$ μs) $- I_{bck}(500+250)$ μs being both the baseline (I_{bck}) corrected photoexcited and unexcited X-ray signals, respectively (see Fig. 1.4). The X-ray signals are measured simultaneously in transmission and fluorescence yield modes, using the same data acquisition strategy for all detectors, as shown in Fig. 1.4. We define the transient signal in fluorescence mode T_{fl} via

$$T_{fl}(E, \Delta t) = \frac{I_{pum}^{fl} - I_{unp}^{fl}}{I_0} \tag{1.6}$$

with I_{pum}^{fl}, I_{unp}^{fl} and I_0 being each baseline corrected as described above for transmission mode. Likewise, normalization of the unpumped spectra to the incident flux yields the static spectrum of the sample.

This data acquisition scheme allows measuring the corresponding signals for every single incident X-ray pulse, and for each data point (e.g., during an energy scan) a few thousand single X-ray pulse intensities are analyzed in order to store the average value and its standard deviation into the computer. This methodology: (i) eliminates drifts due to deterioration of the sample under laser and/or X-ray irradiation (for limited sample volumes), as well as drifts of the X-ray flux; (ii) allows measurements at the shot noise of the X-ray source, implying that changes as low as 10^{-4} X-ray transmission changes can be recorded within a reasonable (typ. 1 s) data acquisition time.

While the above scheme was implemented with 1 kHz repetition rate pump lasers, taking advantage of the high pulse energies available from commercially available amplified laser systems (>1 mJ/pulse), synchrotron X-ray pulses are generally delivered at MHz repetition rates by the storage ring. This implies means that typically 10^3 X-ray pulses are unused. This reduced X-ray flux is a major limiting factor on the achievable signal-to-noise ratio (S/N) of the experiments, which not only reduces the accuracy of the structural analysis but also restricts the types of samples that can be measured to those with a solubility in the range of tens to hundreds of mM, which is not convenient for biological samples. The idea would therefore be to use a pump laser that runs at say, half the repetition rate of the synchrotron but with a large enough energy/pulse to ensure a comparable photolysis yield as in the 1 kHz configuration.

Previous attempts to implement a high-repetition rate pump-probe scheme at synchrotrons were made by Widdra et al. [111, 112] who used a ps laser system operating at 1.25 MHz to match the single-bunch repetition rate at the BESSY synchrotron (Berlin, Germany) for time-resolved core-level photoemission studies of surfaces. Stern et al. [113, 114] used an amplified laser system operating at 272 kHz at the APS to study laser-induced melting of Germanium films. The disadvantage of using high repetition rate excitation with solid samples is the sample damage resulting from the high average power. However, for liquid solutions, high repetition rate excitation is ideal since the flowing sample is continuously refreshed. The Lausanne group recently implemented such an experiment at the SLS using a picosecond high-average power Nd:YVO4 laser operating at variable repetition rate (50 kHz–8 MHz), and producing 10 ps pulses. This product is primarily aimed at industrial users meaning turnkey operation and long-term stability, which are ideal in the environment of a synchrotron. It delivers two orders of magnitude less energy/pulse than 1 kHz amplified femtosecond systems, but the fluence can be identical by tighter focusing of the laser beam. When operated at half the repetition rate of the isolated camshaft pulse of the SLS (520 kHz) this results in 28 μJ/pulse at the fundamental wavelength, 1064 nm. By frequency doubling and tripling one obtains 15 μJ/pulse at 532 nm and 6 μJ/pulse at 355 nm. It is also possible to frequency double the 532 nm light to obtain 266 nm light. The data acquisition scheme is identical to the one described above, with the important difference that the data acquisition rate is increased to 2 MHz. More details are given in [115].

Until recently, most time-resolved XAS studies concerned the hard X-ray range, which is convenient for probing the K-edges of $Z > 18$ elements and the L-edges of $Z > 40$ elements. However, working in the soft X-ray range offers a number of advantages: (a) one could access the K-edges of light elements such as C, N, O, S, etc., which are important for biology; (b) the $L_{2,3}$ edges (due to $2p_{1/2,3/2} \rightarrow 3d$ transitions) of most transition metals are accessed; (c) The smaller intrinsic core hole lifetime width (0.5 eV) of p orbitals results in sharper bound–bound core to valence transitions than K edge ($1s \rightarrow 3d$) ones; (d) $2p_{1/2,3/2} \rightarrow 3d$ transitions are dipole-allowed, yielding more intense and more structured spectra than the dipole-forbidden K edge 1s-3d transitions; (e) the $L_{2,3}$ edge features are directly proportional to the amount of d character of unoccupied or partially occupied valence orbitals of the metal; (f) ligand field multiplet theory [86, 87, 116] is a well-established tool to interpret L edge spectra, delivering a detailed description of the electronic structure.

However, due to the large absorption cross section of solvents and air (e.g., at the Fe L edge energy range the X-ray path length in water is only 1–2 μm) the implementation of soft-X-ray absorption spectroscopy for the study of liquid solutions under vacuum took some time, but is now no longer a problem. Indeed, two strategies have appeared in the last ten to twenty years: (a) the use of cells equipped with soft X-ray transparent windows, such as silicon nitride thin membranes (100–200 nm-thick) [117]; (b) the implementation of the liquid microjet technology, which consists in injecting a liquid jet (typically ≥ 10 μm diameter) at high speed (typically 100 m/s) into the vacuum chamber with a pressure in the order of 10^{-5} bar [118]. The development of ps and fs soft X-ray absorption spectroscopy is very recent and has made use of the first scheme based on SiN-equipped cells [37, 38, 119, 120].

1.4 Applications

The first time-resolved X-ray experiment used X-ray absorption spectroscopy (XAS) to probe the photoinduced changes in carboxymyoglobin in solution with μsec resolution [121]. Later, Wilson and co-workers [97] used 1.5–3 ps X-ray pulses from a plasma source to probe the disappearance of a shape resonance at the K-edge (2.5 keV) of sulphur upon photodissociation of the SF_6 molecule in the gas phase. Chen and co-workers reported the first study of a molecular system in solution, investigating the photodissociation of ligands from [Ni(tpp)L$_2$] (Ni(tpp) = Nickel(ii) tetraphenylporphyrin and L = axial piperidine ligand, which was also the solvent in their study), and their recoordination with 14 ns resolution using synchrotron X-ray pulses. They addressed the issue as to whether the recoordination process is a concerted or a sequential (two-step recombination with [Ni(tpp)L] as an intermediate) process, concluding that the process is concerted on the time scale of their resolution.

Time resolved XAS studies of molecular systems in solution really took off in the first half of the 2000s. Most of these have been transition metal molecular complexes, for obvious technical reason as the methodology was first implemented in the hard X-ray range where it is possible to carry out the experiments in air or under a gas atmosphere. In addition, metal-based molecular complexes, such as metalloporphyrins, haems or chlorophylls, are heavily involved in natural processes involving energy and electron transfer, such as photosynthesis, oxygen transportation, [122, 123] etc., and in applications such as functional molecular devices [124, 125], solar energy, molecular electronics, sensitizers, photocatalysis, etc. Below we give a number of processes occurring in these systems, which have been investigated by time-resolved XAS both in the hard and more recently, in the soft X-ray range.

1.4.1 Intramolecular Charge Transfer

Controlled electron transfer (ET) and charge separation in molecular assemblies are essential requirements for efficient storage and conversion of energy by chemical means. In most light-induced processes in metal-based molecular complexes, the doorway states to excitation are the metal-to-ligand-charge-transfer (MLCT) states, whose absorption bands dominate the visible absorption spectrum. Because a charge is transferred within a molecule there is a significant change of the field of forces within the molecule, but also a dipole moment change. Therefore in solutions, there are issues of both intramolecular and subsequent intermolecular (i.e. solvent)

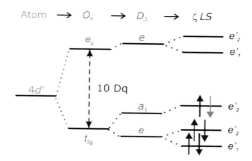

Fig. 1.5 Energy level scheme for a $4d^6$ transition metal complexes such as M(bpy)$_3$, with M = Ru, Fe. The atomic 4d orbital splits into a 6-fold degenerate t_{2g} and a 4-fold degenerate e_g. level in the octahedral field of the ligand (separated by 10Dq). Trigonal distortion due to the D$_3$ symmetry of the complex partial lifts the degeneracy, which is further lifted into 5 two-fold degenerate levels by the 4d spin–orbit coupling. The electron occupancy of the ground state (completely filled in t_{2g}, and empty for e_g) is indicated by the arrows, one (in *red*) is removed to the bpy ligand following photoexcitation of the metal-to-ligand-charge-transfer (MLCT) states

structural rearrangements as a result of the electron transfer. Such processes have been investigated on ruthenium and copper complexes, as described below.

The Lausanne group carried out a picosecond L$_{2,3}$-edge XAS study of aqueous Ruthenium(II)-tris-2,2′-bipyridine ([Ru^{II}(bpy)$_3$]) [88, 126]. In the predominantly octahedral field of the complex, the d-orbitals split into a lower t_{2g} orbital and an upper e_g orbital (Fig. 1.5).

These are further split by the addition of the trigonal distortion and by the spin–orbit interaction. Fig. 1.6 shows the ground state (a) and transient XAS (b) spectra, next to the extracted excited state XAS (c). In the ground state, all six 4d electrons are in the lower t_{2g} orbital while the e_g orbital is empty, and therefore only the B band (2p → 4d(e_g)) appears. In the MLCT state, an electron is transferred to the bpy ligand, thus opening up a new transition (2p → 4d(t_{2g}), feature A'). In room temperature aqueous solutions, the ^{3}MLCT state exhibits a measured lifetime of about 600 ns [127]. Also, the blue-shift of the B feature reflects the change of oxidation state, in good agreement with previous experimental and theoretical work [128, 129].

The XANES region was analyzed by the ligand field multiplet theory, [88] in very good agreement with the data (see Fig. 1.6d). From the change of ligand field splitting between ground and excited state and using an electrostatic model that relates the octahedral ligand field splitting to the metal–ligand distance, [130] a Ru–N bond contraction of ~0.02 Å in the excited state was derived, while the analysis of the EXAFS region [64] (D, D′ features) delivers a value of the Ru–N bond contraction of -0.037 ± 0.0135 Å, treating all Ru–N distances equally (i.e., in D$_3$ symmetry). This relatively weak bond contraction, despite a dramatic change of the electronic structure, results from steric effects because the three byp ligands are already in a constrained geometry in the ground state. Our results were later confirmed by Quantum chemical calculations of the ^{3}MLCT structure by Nozaki et al. [131] and Alary et al. [132].

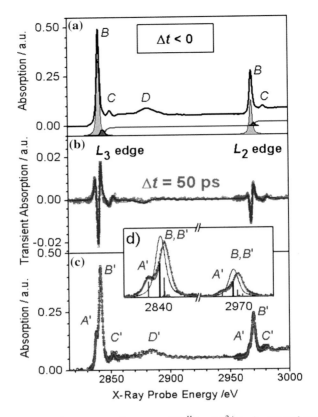

Fig. 1.6 a $L_{2,3}$ edge XAS spectrum of aqueous $[Ru^{II}(bpy)_3]^{2+}$ in its ground state (*thick black line*) together with fits of its most prominent features (labeled), except for the two residual bands (*green areas*) near the ionization potential. The edge step functions for IP for each state is shown as well (*blue line*). See [88] for details of the fit. **b** Transient absorption spectrum (*open red circles* with error bars) and a fit of this transient (*blue line*) using the ground state spectrum and the (energy-shifted) decomposed bands shown in **a**. **c** Reconstructed XAS of the ^{3}MLCT state (*blue data points*) together with a fit to the most prominent features. **d** The A(A′) and B(B′) features extracted from the spectrum after subtraction of the edge step and the higher-lying weak bound–bound absorptions shown in **a**, and compared to multiplet calculations. The sticks (thick for the ground state and thin for the excited state) are the transitions which are broadened with the Gaussian and Lorentzian widths due to experimental resolution and lifetime width, to generate the solid (*red*) curves

In another example of structural changes due to intramolecular electron transfer processes, Chen et al. [133] investigated the structure of the MLCT states of Cu(I)-diimine complexes ($[Cu^{I}(dmp)_2]^+$ (dmp = 2,9-dimethyl-1,10-phenanthroline)) by ps XAS. Upon 400 nm photoexcitation of the system one reaches the S_2 state, which decays on an ultrafast time scale to the S_1 state [133–136] that then relaxes to the T_1 state (Fig. 1.7a). In these processes, the system changes from a tetrahedral geometry to a flattened one (Fig. 1.7b).

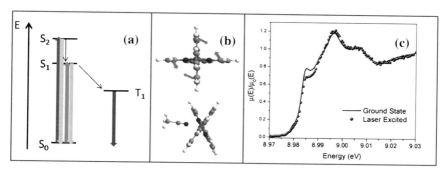

Fig. 1.7 Photocycle of cooper(I)-diimine complexes: upon excitation of the S_2 MLCT state the system changes from Cu(I) to Cu(II) and an ultrafast internal S_2-S_1 conversion occurs (**a**). In this step, a geometry change towards a flattening occurs (**b**), which is thought to take place in the S_1 state prior to intersystem crossing to the T_1 state [133, 134, 208]. The latter has a ns lifetime so that it can be probed by ps XAS as done in refs. [136, 209] (**c**)

Chen et al. observed the photoinduced electronic and structural changes by ps XANES and EXAFS (Fig. 1.7c) and found that: (1) the excitation of the MLCT state transition induces an electronic configuration change from Cu^I ($3d^{10}$) to Cu^{II} ($3d^9$); (2) the inner-sphere reorganization changed the coordination number of the MLCT state from four to five in toluene, which is presumed to be non-coordinating; (3) the average Cu-ligand bond lengths increased in the MLCT state in toluene, but decreased in acetonitrile, reflecting the difference in the interactions of the copper with the fifth ligand. These conclusions were further supported by simulations of the XANES part of the spectra [137]. In the meantime, we have carried out ps XAS measurements in similar complexes and complemented them by Quantum Mechanics/Molecular Mechanics molecular dynamics simulations, and came to the conclusion that there is no coordination of a solvent molecule to the Cu atom [138]. Nevertheless, questions still remain open as to the sub-ps dynamics, which may be addressed by fs XAS in the future.

1.4.2 Bond Formation in Bimetallic Complexes

The triplet excited states of dinuclear d^8--d^8 platinum, rhodium, and iridium complexes (bridged by various ligands) exhibit remarkable photophysical and photochemical properties, which are strongly determined by their structure [140]. The unusually high photocatalytic activity of these complexes are a manifestation of the newly formed bond in the lowest excited singlet and triplet $^{1,3}A_{2u}$ states, owing to the promotion of an electron from the antibonding $d\sigma^*$ (d_z^2-derived) to the bonding $p\sigma$ (p_z-derived) orbitals, which should therefore lead to a contraction of the metal–metal bond (Fig. 1.8).

The $[Pt_2(P_2O_5H_2)_4]^{4-}$ molecule is among the most extensively studied d^8--d^8 dinuclear metal complexes. In solutions, excitation into the first singlet state in

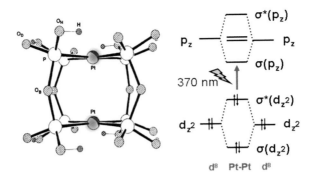

Fig. 1.8 Structure of the $[Pt_2(P_2O_5H_2)_4]^{4-}$ molecule (*left*). The maximum of the first singlet absorption lies at 370 nm. Excitation of a d^8–d^8 complex promotes an electron from the antibonding $d\sigma^*$ (d_z^2-derived) to the bonding $p\sigma$ (p_z-derived) orbitals, leading to formation of the Pt–Pt bond and contraction of its distance in the excited state

the near UV region around 370 nm leads to formation of the long lived (~1 μs) triplet state with unity quantum yield. van der Veen et al. [141] resolved its structure by picosecond EXAFS. Figure 1.9a shows the ground state Pt L$_3$-edge XAS spectrum (black trace) as well as the transient spectrum, integrated from 0 to 150 ns to improve the signal-to-noise ratio. The inset shows the XANES region for the ground-state complex and its transient spectrum, wherein dramatic changes appear. In particular, a new absorption shows up at 11.574 keV below the absorption edge, which is due to the creation of a hole in the $5d\sigma^*$ orbital upon laser excitation, which can then be accessed from the $2p_{3/2}$ core orbital (L$_3$ edge). Clear changes are visible in the EXAFS region (Fig. 1.9b), reflecting structural modifications between the ground and excited triplet states.

From the transient EXAFS spectrum (Fig. 1.9b), the magnitude of the Pt–Pt bond contraction as well as, for the first time, the changes affecting the Pt–P bonds were extracted, [141] and the best fit is shown in Fig. 1.9b. It was found that while the Pt–Pt bond contracts by 0.31(5) Å, in very good agreement with the X-ray photocrystallography data [142], it was shown that Pt–P bonds slightly elongate (by ~20 mÅ) in agreement with theoretical predictions [142]. This work underscores the ability to retrieve details of the excited state structure of a rather complex molecular system in liquid solution, due to the high sensitivity of the experiment, coupled to a rigorous structural analysis based on fitting the transient EXAFS spectra directly in energy space [83]. Almost simultaneously to the work by van der Veen et al. [141], Christensen et al. [27] reported a picosecond solution X-ray scattering study of the same system, deriving a somewhat similar contraction of the Pt–Pt bond. More recently, Chen and co-workers [143] reported on ps XAS studies of another diplatinum complex, which also produced rather similar results.

Pt$_2$POP is a remarkable molecule in the sense that its ground and excited potentials are highly harmonic and that the optical spectrum is only sensitive to the Pt–Pt coordinate. It is therefore easy to retrieve the excited state structure from a Franck–Condon analysis, which shows a Pt–Pt bond contraction in the triplet

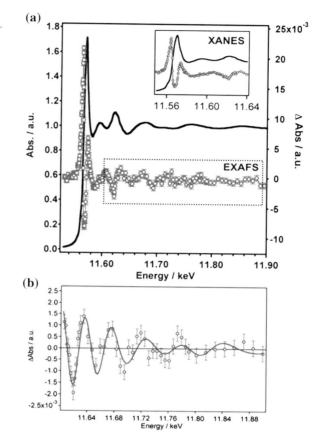

Fig. 1.9 **a** Static Pt L$_3$ XAS spectrum of [Pt$_2$(P$_2$O$_5$H$_2$)$_4$]$^{4-}$ in solution (*black line*, *left* axis) and the transient (excited–unexcited) XAS spectrum (*red circles, right* axis, same units as *left*) integrated up to 150 ns after excitation. The inset zooms into the XANES *region*. **b** Transient EXAFS data and best fit (*solid line*) with the following results: a Pt–Pt contraction of 0.31(5) Å, a Pt-ligand elongation of 0.010(6) Å, zero energy shift and 7 % excitation yield. The best-fit structural distortions are indicated in the upper right corner (see [141, 206] for details)

state 0.1 Å smaller than that derived in [141–144]. The origin of the discrepancy lies, we believe, in the fact that the EXAFS analysis neglected the contribution of the solvent, while the optical domain spectroscopy is not sensitive to it. This shows that solvent contributions have to be included in the analysis of EXAFS and XANES, which opens a new level of refinement in the structural determination. Work is in progress towards such refinements. A recent paper reports on such calculations and show a clear contribution of the solvent in the signals [139].

1.4.3 Spin Cross-over in Fe(II)-Complexes

One of the fascinating features of Fe^{2+} or Fe^{3+} metal-based molecular complexes is their ability to change spin under temperature, pressure or light irradiation and they have therefore been named spin cross-over complexes (SCO) [145]. In the

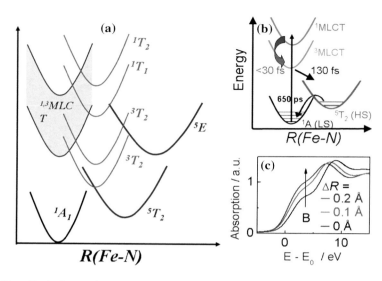

Fig. 1.10 **a** Typical potential energy curves of Fe(II)-complexes as a function of the Fe–N bond distance adapted from [146]. The manifold of metal-to-ligand-charge-transfer (MLCT) states is shown as a shaded area. $[Fe^{II}(bpy)_3]^{2+}$ has a predominantly O_h symmetry with a trigonal (D_3) distortion. The metal-centred (MC) states are represented by their symmetry character (A, T and E) in the D_3 group. In the latter, the LS $^1A_1$ ground state has a completely filled $e^4a_1^2$ configuration (deriving from the t_{2g}^6 subshell in O_h symmetry), while the antibonding e (e.g. in O_h symmetry) orbital is empty. Per electron promoted from the t_{2g} to the $e_{.g.}$ subshell (for easier reading we will use from here on the O_h nomenclature) the metal–ligand bond length increases by as much as 0.1 Å [146]. For the series of $^{1,3}T(t_{2g}^5e_g)$ states, it is expected to be in between the ground and the high-spin $^5T_2$ ($t_{2g}^4e_g^2$) state. **b** Relaxation cascade as determined by ultrafast laser spectroscopy upon excitation of aqueous $[Fe(bpy)_3]^{2+}$ at 400 nm [156]. The intermediate MC states are not shown as they are optically silent in the UV–Visible and were therefore not observeable. **c** Simulated XANES spectra for different Fe–N bond elongations from the ground state equilibrium value of 1.97 Å, using the MXAN code [91]. The elongations correspond to the ground and the MLCT states (0 Å), the intermediate ligand field $^{1,3}T$ states (0.1 Å) and the ^5T state (0.2 Å). The horizontal axis is given as energy from the ionization potential

predominantly octahedral field due to the ligands (Fig. 1.5), all electrons are in the lower t_{2g} sub-shell in the low spin (LS) ground state, while transferring electrons to the e.g. orbitals increases the spin state. Because the e.g. orbitals derive from the d_{x-y}^2 and d_z^2 orbitals, they are antibonding in 6-fold coordinated complexes, which leads to a striking metal–ligand bond elongation in the high spin (HS) state. A generic diagramme of the potential energy curves of the various states of Fe(II)-based complexes is shown in Fig. 1.10a, as a function of the Fe–N bond length [146]. The MLCT states have nearly the same equilibrium distance as the ground state, in agreement with [88], the ligand field states $^{1,3}T$, ^5T and ^5E have their equilibrium distances elongated by ~0.1, ~0.2 and ~0.3 Å, respectively, relative to the ground state bond distance. Light excitation into the singlet Metal-to-Ligand-Charge-Transfer (^{1}MLCT) state or to the lower-lying ligand field states leads to

Fig. 1.11 **a** XANES
spectrum of the LS state of
[Fe^{II}(bpy)$_3$]$^{2+}$ (*black*) and
the HS state (*red points*).
The latter is extracted from
the difference spectrum
(shown in **b**) and the LS
spectrum, based on the prior
determination of the fraction
of excited molecules.
b Difference transient
spectrum at 50 ps time delay
(*red points*), and at 300 fs
(*blue stars*) obtained with the
slicing scheme

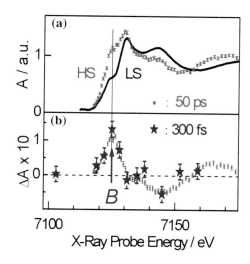

population of the lowest quintet state $^5T_2$ with unity quantum yield [145]. The lifetime of the latter varies by several orders of magnitude as a function of ligand and temperature [145] with [Fe^{II}(bpy)$_3$] having the shortest lived quintet state lifetime (650 ps) at room temperature.

Structural studies by X-ray diffraction or X-ray absorption spectroscopy under quasi steady-state conditions, pointed to a bond elongation of ~0.2 Å of the Fe–N bond for Fe(II)-based complexes with long-lived HS states [147–149] but the question arose if this applies to the shortest lived HS state of [Fe^{II}(bpy)$_3$], as predicted by theory [150].

Khalil et al. [151] and Gawelda et al. [152] captured the structure of the quintet state after laser excitation using 70 ps hard X-ray pulses probing the structure changes at the K edge of Iron. In the first case, the system was [Fe^{II}(tren(py))$_3$] in acetonitrile, which has a quintet state lifetime of 60 ns, while the second case was [Fe^{II}(bpy)$_3$], which we describe in more details now. Figure 1.11 shows the Fe *K*-edge XANES of the molecule in the ground state, the transient (difference) spectrum at 50 ps time delay and the XAS spectrum of the quintet state, as retrieved from the ground state and the difference spectra and from the photolysis yield determined in laser-only experiments (and convoluted to match the much longer X-ray probe width). The structural analysis of the excited state was based on fitting both the transient XANES [152] and the transient EXAFS spectra [58, 83] in energy space. These deliver the same Fe–N bond elongation $\Delta R_{Fe-N} = 0.20$ Å, but the precision increased considerably using the latter procedure (0.203 ± 0.008 Å).

The fact that the bond elongation is nearly the same in the HS state in all complexes, regardless of its lifetime [147, 151] implies that the bond elongation is not the parameter that controls the decay rate of the quintet state to the ground state. Rather, the adiabatic energy and the coupling parameters between low spin and high spin state are the crucial parameters. Indeed, of all Fe(II)-SCO complexes, $[Fe^{II}(bpy)_3]^{2+}$ has the highest lying quintet state.

In order to describe the electronic structure of the quintet state, Huse et al. [120] carried out picosecond L-edge spectroscopy of $[Fe^{II}(tren(py))_3]$. This experiment was the first of its sort and it used a liquid cell equipped with SiN membrane windows. Their results showed a significant reduction in orbital overlap between the central Fe(3d) and the ligand N(2p) orbitals, consistent with the expected ~0.2 Å increase in Fe–N bond length upon formation of the high-spin state. The overall occupancy of the Fe(3d) orbitals remains constant upon spin cross-over, suggesting that the reduction in σ-donation is compensated by significant attenuation of π-back-bonding in the metal–ligand interactions. These results show the power of ps soft X-ray spectroscopy to fully unravel the details of the electronic structure of the systems under study, and its complementarity with hard X-ray spectroscopy.

Femtosecond resolution in X-ray absorption spectroscopy is however needed if one wants to follow the ultrafast structural dynamics of photoexcited systems in real-time, and thus answer a host of questions that have escaped femtosecond optical spectroscopies or ps XAS. Femtosecond XAS was recently implemented in the hard and the soft X-ray ranges to fully resolve the photocycle of the above spin cross-over systems.

Indeed, the mechanism and relaxation pathways of the photoinduced HS-LS conversion in Fe(II) complexes were still unknown, and in particular the pathway leading from the initially excited ^{1}MLCT state to the lowest lying excited quintet state [145, 153, 154]. One of the reasons for this situation was the fact that the intermediate ^{1,3}T state (Fig. 1.10a) are optically silent, while the HS quintet state absorbs below 320 nm [155]. The early time dynamics was identified by Gawelda et al. on $[Fe^{II}(bpy)_3]$ in laser-only studies, [156] who found that departure from the ^{3}MLCT state occurs in ~130 fs (Fig. 1.10b). The XANES spectrum of the ground state (Fig. 1.11) shows a shoulder (the B-feature) on the edge, which is a multiple scattering feature [157]. Upon Fe–N bond elongation in the quintet state its intensity increases causing the significant peak at low energy in the transient spectrum (Fig. 1.11b). Simulations based on multiple scattering theories confirm the distance dependence of the B-feature (Fig. 1.10c), which one can use to identify the various electronic states possibly involved in the relaxation cascade.

Using hard X-ray fs pulses generated by the slicing scheme at the SLS, Bressler et al. carried out an optical pump/X-ray probe experiment following the evolution of the B-feature as a function of pump-probe delay. Figure 1.12 shows the time scan obtained at the B-feature (7.122 keV) showing that the signal stabilizes from about 300 fs up to the scan limit of 10 ps (see inset), which is evidence that the system is already in the quintet state. Further evidence is the energy scan at 300 fs, which reproduces the transient spectrum at 50 ps time resolution (Fig. 1.11b). The fit in Fig. 1.12 shows that the quintet state is reached in 150 ± 50 fs. This time corresponds to the decay of the ^{3}MLCT state, implying that the ultrafast spin conversion is a simple three-step ^{1}MLCT - ^{3}MLCT - $^5T_2$ cascade that bypasses the intermediate ^{1,3}T states.

The time scale of ~150 fs corresponds to about two oscillations of the Fe–N stretch vibration [158], suggesting a nearly non-Born–Oppenheimer process. This

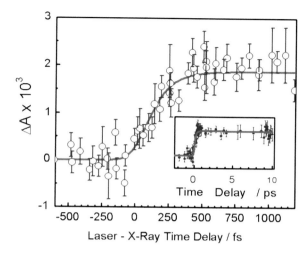

Fig. 1.12 Time scan of the signal (*blue points*) at the B-feature (Fig. 1.6c), as a function of laser pump/X-ray probe time delay after excitation of aqueous $[Fe^{II}(bpy)_3]^{2+}$ at 400 nm. The inset shows a long time scan up to 10 ps time delay. The *red trace* is the simulated signal assuming a simple 4-step kinetic model $^1A_1 \rightarrow {}^1MLCT \rightarrow {}^3MLCT \rightarrow {}^5T$ to describe the spin conversion process. The vertical arrow displays the expected absorption increase for an elongation of 0.2 Å for the Fe–N distance ΔR between the LS and HS states (see Fig. 1.10c)

experiment illustrates the power of ultrafast XAS to retrieve dynamical information difficult to obtain by laser-only experiments. It also shows how structural dynamics studies can identify the electronic relaxation pathways of complex molecules, while so far the opposite was common practice. Further visible pump/UV probe transient experiments of the quintet state identified its vibrational relaxation dynamics, [159] thus providing a complete picture of the photocycle of aqueous $[Fe^{II}(bpy)_3]$. A full description of the combined optical and X-ray studies on this system is given in a recent review [160].

Very recently, Huse et al. [38] carried out the first time femtosecond soft X-ray experiment of a similar spin cross-over molecular system in solution, which fully confirm the above results. Their experiment represents a real breakthrough as it demonstrated the visualization of ultrafast electronic structure changes of dilute molecular systems in solution by soft XAS.

1.4.4 Dynamics of Pure Water

The unusual properties of water have stirred much research. Liquid water can be regarded as a highly dynamic network of molecules connected through hydrogen bonds. The results of static X-ray absorption spectroscopy measurements show a pre-edge peak (I) followed by a main-edge (II) and a postedge feature

(III) related to local water structures with specific coordination and hydrogen bonding. There has been much controversy over the correct structural model and their relative merits in determining the average number of hydrogen bonds per water molecule [117]. Nevertheless, there is consensus on a qualitative level as to the structural meaning of the three characteristic spectral regions of the oxygen K-edge and how these features change upon melting of ice as well as heating of liquid water. In order to describe how the different structures interchange upon heating of water, Huse et al. [161] and Wernet and co-workers [119, 162] demonstrated infrared pump/soft X-ray probe investigation of pure water at the K edge of Oxygen. However, the tens of ps time resolution of these experiments precluded observations at relevant time scales. Therefore, Wen et al. [37] extended these experiments to the femtosecond range using the slicing scheme at the ALS. They recorded the conversion of highly coordinated water structures to less-ordered structures with weaker hydrogen-bonding via fs-XANES spectroscopy upon vibrational excitation, exploiting the relation between coordination/order and distinct X-ray spectral features due to the high sensitivity of XANES regions to chemical coordination. The observed dynamics was described by a 0.7 ps thermalization time that is characteristic of the hydrogen bond network in water. Their transients capture the formation of a high-pressure phase distinct from the equilibrium state of liquid water.

This work was the first femtosecond soft X-ray study of a liquid-phase system. It demonstrated the feasibility and prospects of solution-phase ultrafast X-ray studies in the soft X-ray range where K-edges of C, N, O and L-edges of transition metals deliver chemically specific information about coordination, valence charges distribution, bonding and atomic structure in order to uniquely follow chemical reactions and solvation dynamics on ultrafast time scales. As a matter of fact, the same team demonstrated ps and fs soft X-ray spectroscopy at the L edge of Fe in an SCO complex as discussed above ([38, 120]) and at the N K-edge of photoexcited [$Fe^{II}(bpy)_3$], [120].

1.4.5 Solvation Dynamics and Hydrophobicity

The above examples on Copper, Platinum and Ruthenium complexes alluded to the role of the solvent in the photoinduced structural changes. Electronic solvation dynamics is the rearrangement of the solvent molecules around a solute as a result of an electronic redistribution in the latter. This electronic redistribution may be caused by photoexcitation or simply by a chemical reaction. The role of the solvent is crucial in enhancing or hindering reactions, via its modification of barriers and the dynamics of solvation is intimately influenced by the intramolecular dynamics. So far solvation dynamics has only been investigated by ultrafast optical techniques (transient absorption or fluorescence up-conversion), [163–165] which do not deliver structural information. In these studies a dye molecule is excited in a solvent, inducing either a dramatic dipole moment change

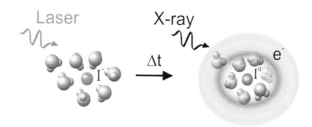

Fig. 1.13 Principle of the optical pump/X-ray probe experiment on solvation dynamics. The electron of the iodide (isoelectronic with Xe) is removed by an ultrashort laser pulse. The evolution of the structural changes in the solvation shell is probed by an X-ray pulse that records the changes at the iodine L edges

(polar solvation) or an expansion of the electronic cloud (non-polar solvation) in such a way that depending on the polar or non-polar nature of the solvent, respectively, a dramatic solvent shell reorganisation occurs, which minimises the free energy. While this approach delivers detailed information about the dynamics of the system, the solvent shell structure cannot be recovered. In addition, the solvent rearrangement depends on the shape and size of the molecule in the ground state. In order to capture the solvent structural dynamics and to obtain a general description of the solvation dynamics, we carried out time-resolved XAS studies of the solvent rearrangement around photoexcited aqueous iodide. In this case, the UV (<260 nm) photoexcitation abstracts the electron from the iodide, leaving a neutral iodine behind (Fig. 1.13) and the solvated electron. It turns out that this changes the solute from a hydrophilic one to a hydrophobic one.

Hydrophobicity is considered the major driving force behind fundamental biological and chemical processes. In the hydration of small hydrophobic species, the formation of small cavities in the solvent to accommodate the solute is an entropically dominated process, and the presence of the solute constrains the orientational and translational degrees of freedom of the neighbouring water molecules. The microscopic understanding of hydrophobic hydration relies so far almost solely on theory and simulations, due the fact that experimentally probing hydrophobicity at the atomic scale is very difficult, and quantitative studies are not yet available.

We previously carried out a preliminary picosecond XAS study of multiphoton-excited aqueous iodide, which showed significant changes in the XANES and EXAFS regions of the $L_{1,3}$ edges 50 ps after excitation, suggesting an extensive solvent rearrangement in the transition from iodide to iodine [166]. Soon after, Elles et al. [167] reported a similar study on aqueous bromide excited into the so-called charge-transfer-to-solvent (CTTS) bands by one-photon at 200 nm, and probed the resulting neutral bromine atom was done at its K-edge using picosecond hard X-ray pulses. Their recovered Br^0 K-edge spectrum showed shallow features above the edge, making it difficult to extract a solvent shell structure. For this purpose, they carried out Monte-Carlo simulations of the radial distribution functions (RDF) of Br^- and Br^0, and found that indeed in the latter case,

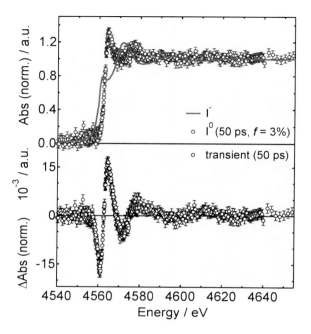

Fig. 1.14 a Normalized static iodide L_3 edge XAS (*solid line*) and reconstructed I^0 spectrum using the transient spectrum at 50 ps time delay **b** and the photolysis yield of 3 % (see 39), sample concentration 100 mM; **b** Normalized L_3 edge transient spectrum at a time delays of 50 ps

the solvent shell structure is more diffuse and expands by ~0.5 Å, explaining this way the weak modulations of the spectrum above the Br^0 K-edge. Their RDF also exhibited a small peak at shorter distances, which they attributed to formation of a $Br^0...OH_2$ charge transfer complex, although no signature of this complex was found in the time-resolved X-ray signal.

Pham et al. [39] recently carried out a detailed ps and fs XAS study of photoexcited aqueous iodide, which was complemented by detailed quantum chemical calculations and molecular dynamics simulations [39]. While the L_1-edge (originating from the 2 s core orbital) informs us about the electronic structure of the systems, in particular via the appearance of the 2 s-5p resonance that becomes possible upon formation of neutral iodine, the L_3 edge (originating from the $2p_{3/2}$ core orbital) is ideal for probing the geometric structure of the solvent shell. Figure 1.14 shows the static L_3 iodide spectrum (solid line in (a)), the transient spectrum (b) and the recovered iodine spectrum (points in (a)). Dramatic changes are observed which point to an extensive rearrangement of the solvent shell. In order to quantify it, we carried out both classical and quantum mechanics/molecular mechanics (QM/MM) molecular dynamics (MD) simulations. The iodide solvent shell had already been described by QM/MM MD simulations, [168] and the resulting RDF is show in Fig. 1.15 for both the I–O and the I–H distances. Figure 1.15 also shows the RDF's obtained by classical and QM/MM MD for iodine. Both types of MD simulations show an increase of the I–O distances from iodide to iodine, but the most dramatic changes show up on the I–H RDF, where the first peak shifts on average by well over 1 Å

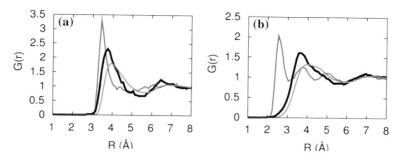

Fig. 1.15 I-O (**a**) and I-H (**b**) radial distribution functions (RDF) of aqueous iodide (*red lines*, same as in [168]) and iodine for the classical MD (*green lines*) and the QM/MM MD (*black lines*). The QM/MM MD simulations represent an equilibration over 15 ps with a single I^-/I^0 described at DFT level of theory. The I^0-O/H RDFs were equilibrated for 10 ns in the CMD simulations

from iodide to iodine. While for iodide the first peak of the I–H RDF lied at smaller distances than that of the I–O RDF, for iodine it is at almost the same distance as the first peak of the I–O RDF, but it is broader and has a long tail extending to larger distances. Finally, the I–H RDF shows clear modulations in the case of iodide, but the separation between solvent shells is less clear cut in the iodine case.

All this indicates that the majority of the hydrogen atoms rotate and point away from the iodine atom, contrary to the iodide case. We used these RDFs to simulate the transient L_3 EXAFS spectrum and both classical and QM/MM MD of neutral iodine delivered a good agreement with the experimental data. The main result is that the experimental data confirms the expansion of the solvent cage from I^- to I^0, with an increase of the cage radius (measured on the I–O distance) of 5–20 % (depending on the type of simulations). Furthermore the RDFs already suggest the formation of a hydrophobic cavity as the water hydrogen atoms prefer now to point towards other water molecules rather than towards the solute.

Sub-picosecond transient XANES spectra are shown in Fig. 1.16 where one can note a broadening on the blue side of the features that appear at the L_1 edge, compared to the 50 ps transient. Based on quantum chemical calculations and QM/MM MD simulations of an $I^0(H_2O)_9$ cluster in bulk classical water, this broadening is caused by the occurrence of a weakly bound and short lived (few ps) I^0-OH_2 complex, whose 2 s-5p resonance is slightly blue shifted compared to that of the uncomplexed Iodine atom due to the formation of a 3-electron bond between iodine and water, therefore causing the observed broadening.

The simulations show that the lifetime of the complex is determined by the time it takes to complete the formation of a cavity of hydrogen bonded molecules around the solute (hydrophobic solvation) following electron abstraction from iodide. This study showed the power of combining ps and fs XAS with high-level quantum calculations and molecular dynamics simulations to unravel the complete dynamics of the system.

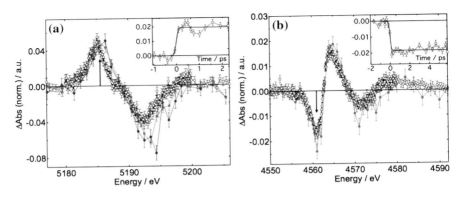

Fig. 1.16 a Normalized transient L_1 edge spectrum of photoexcited aqueous iodide at 50 ps (*black dots*), 300 fs (*red dots*) and 850 fs (*blue dots*) time delays after excitation. **b** same as **a** but for the transient L_3 edge spectrum. The sub-ps transients were multiplied by a factor of 1.75 in order to match the value of the 50 ps transient at the maximum of the positive feature. The insets show the time traces recorded with fs resolution at 5185.5 eV **a** and 4561 eV **b**, denoted by arrows, together with their fits by a step function with a rise time of 250 fs corresponding to the cross correlation of the optical and X-ray pulses

1.4.6 Towards Biological Systems

Metalloporphyrins are the basic constituents of a large class of biological systems and play a central role in photosynthesis and in respiration [169]. Time-resolved XAS should solve a number of important questions concerning their relaxation pathways and mechanism after photoexcitation. Indeed, similar to the above presented Iron(II)-polypyridine complexes, metalloporphyrins contain several optically silent metal-centred (or so-called *d–d* states), which can be accessible by XAS.

Chen et al. pioneered the study of metalloporphyrins in solution [55, 59, 170–172]. They first studied the ligand (L) photodissociation of [Ni(tpp)L$_2$] (Ni(tpp) = Nickel(ii) tetraphenylporphyrin and L = axial piperidine ligand, which was also the solvent in their study), and their recoordination with 14 ns resolution using synchrotron X-ray pulses [171]. The process has a time constant of 28 ns, which is sufficiently long to be resolved. They demonstrated time-resolved XAS measurements for dilutions as low as 1 mM, which are important for biology. They also found that the photodissociation intermediate is square-planar under their time resolution limit. Next, they investigated the structure of the unligated nickeltetramesitylporphyrin (Ni^{II}TMP) in the excited triplet state. Here the idea was to resolve the transient electronic structure. Indeed, The electronic configuration of Ni(II) ($3d^8$) in an almost square-planar ground state S_0 has an empty $3dx^2\text{-}y^2$ molecular orbital (MO) and a doubly occupied $3dz^2$ MO. Upon $S_0 \rightarrow S_1$ transition, the S_1 state is believed to decay to an intermediate state T_1' state that then undergoes vibrational relaxation to a relaxed T_1 triplet state, with a presumed

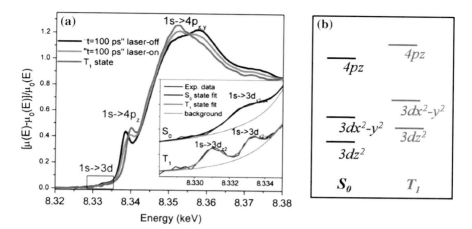

Fig. 1.17 **a** Ni K-edge XANES spectra of NiTMP; inset: pre-edge region. *Black*, laser-on at t = 100 ps; *green*, laser-on at t = 100 ps; and *red*, the T1 state spectrum; t = delay time). The X-ray absorption near-edge structure (XANES) spectra of the Ni K-edge is rotated counter-clockwise by 90° from the conventional display with highlighted regions enlarged showing the pre-edge region for the 1 s → 3d transitions, and the transition edge region for the 1 s → 4pz transitions

^{3}(3dx^2-y^2, 3dz^2) configuration, where 3dx^2-y^2 and 3dz^2 MOs are each singly occupied. The T$_1$ state returns to the ground state in approximately 200 ps.

By ps XANES (Fig. 1.17a), Chen et al. found a single peak appears in the pre-edge region due to the 1 s → 3dx^2-y^2 transition in the S$_0$ state, while two peaks due to the 1 s → 3dx^2-y^2 and 1 s → 3dz^2 transitions are observed in the excited T$_1$ state, showing for the first time that indeed the latter has two singly occupied 3dx^2-y^2 and 3dz^2 MOs (Fig. 1.17b), which was only know from quantum chemical calculations. These results show the power of time-resolved XAS at unraveling the electronic structure of transient species that are optically silent. In addition, from the EXAFS they also obtained the geometry changes of the excited system, as well as following the ligation processes that occur with some solvents [173].

The next stage along this line of results is the study of metalloproteins and in particular, haemoproteins. As a matter of fact the first time-resolved XAS ever to be carried out were on ligated myoglobins, [121, 174–179] but were all limited to the microsecond time scale. This time resolution is sufficient for the CO ligand, and for other ligands if the system is at low temperature, which was the case in some of these studies. To get at the dynamics at room temperature, one would need to reach the ps, and possibly the fs tme resolution [180–183]. In order to reach the ps resolution and exploiting its enhanced sensitivity, the Lausanne group implemented the high repetition rate scheme described in Sect. 1.2 and [115]. A first result on the photodissociation of CO from carboxymyoglobin (MbCO) is shown in Fig. 1.18. Because of the long recombination time of CO to Mb (milliseconds), this results was not obtained with the pulse-to-pulse data acquisition

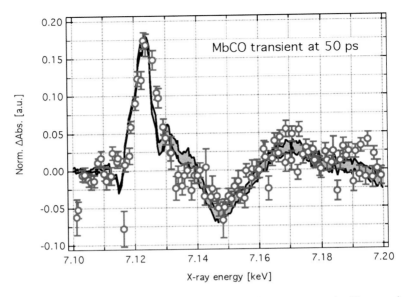

Fig. 1.18 Transient XAS of 2 mM MbCO collected in fluorescence mode. The sample was excited at 532 nm. The black curve represents the difference XAS based on the static spectra of deoxyMb and MbCO and scaled by 21 % (after [115])

described in Sect. 1.2. Rather the spectra were accumulated at tens of ps time delay with and without the laser on. Figure 1.18 shows that the transient spectrum is similar to the difference of the ligated minus unligated (deoxy form) spectra of Myoglobin, confirming that the data acquisition strategy is sound. This study was then extended to probing the recombination dynamics of MbNO, whose time scale is about 200 ps. This was done using the pulse-to-pulse data acquisition and the results fully confirm the recombination time derived from optical-only studies [184]. These first results confirm that ps XAS can be implemented on biological systems and open the way to the study of ligand dynamics with time resolution of better than 100 ps. They are also important forgoing studies for the preparation of experiments to be carried out at the X-ray free electron laser.

1.5 Conclusion and Outlook

The above review presented the recent advances in picosecond and femtosecond X-ray absorption spectroscopy but it is not exhaustive. The more recent activities of the Argonne group on molecular systems in liquids have been reviewed in [59]. This review also does not include the numerous ps and fs XAS studies carried out on solid materials, which have led to a wealth of new results in condensed matter physics on highly correlated systems, [42, 185–187] on semiconductors, [188] on phase transitions, [100, 101, 189–192] defect formation in solids, [193] and on the spin dynamics in materials [194].

Ps XAS (and to a lesser extent fs XAS) can be considered as a routine technique that can be implemented to a large class of molecular systems in solution, but also to the probing of atomic ions in solutions, as recently demonstrated in the case of aqueous Iodide [166] and Bromide [167]. These studies presented above, have been limited to the systems containing heavy atoms, since the solvent absorption is weak in the hard X-ray range. Furthermore, they have been limited to chemical systems, even though the desire to carry out studies on biological systems in physiological media has been stressed in the introduction to the technique, and as already mentioned, the first time-resolved XAS was carried out on biological systems [121]. Such studies are still commonly being carried out with milli- to microsecond resolution [148, 149, 170, 178, 179, 195, 196]. One of the aims of current efforts is to extend them to higher temporal resolutions.

Several exciting prospects are on the way thanks to recent developments in the implementation of methodologies:

a. All the above experiments were carried out using a pump laser at 1 kHz repetition rate, while the storage ring operates at MHz repetition rates. This implies that typically 10^3 X-ray photons remain unused. The implementation of high repetition rate lasers (running at hundreds of kHz) that match an integer fraction of the storage ring repetition rate, combined with microfocussing of the X-ray and laser beams (this is possible at several synchrotron beam lines) allow a significant increase of signal to noise, and therefore, the possibility to reduce the data acquisition times and to work at much lower dilutions that in the present situation (tens of millimolar), thus approaching the concentration of biological samples [115]. In addition the compactness of the high repetition rate laser systems allow for a large flexibility in portability and simplification of the experimental set-up. Last but not least, the use of the high repetition rate laser allows for new experiments such as X-ray emission (XES), [197] resonant inelastic X-ray scattering (RIXS) and X-ray diffraction (XRD) and scattering (XRS). As a matter of fact, recent work by Bressler et al. [198] have indeed shown all these capabilities.

b. Now that femtosecond XAS been demonstrated [36, 38, 39] thanks to the slicing scheme, more studies should be possible. However the main limitation of the slicing scheme is its very low flux, and more hope is now put on the free electron lasers which deliver orders of magnitude more flux. Recently, femtosecond resolved XAS spectra at the Fe K edge were recorded at a commissioning beamtime at the LCLS Stanford, demonstrating very short data acquisition times [199].

c. Extension of time-resolved XAS to the soft X-ray regime [57, 58], and the recent results by Huse et al., both in the ps [119, 120] as well as the fs time domain [37], show the anticipated returns from such a technique. The recent static soft X-ray studies of biological systems in physiological solutions [200] also open the way to their study in the time domain. Furthermore, the recent demonstration of a strong interaction of aqueous solvents with metal centres in various types of complexes, offers a new way of probing the role of the solvents in affecting the electronic structure of solutes, [201] but here again, when transferred to the time-domain these studies will deliver an additional degree of insight. These approaches have already yielded a wealth of new results on chemical systems.

d. Core-level spectroscopies, such as Auger spectroscopy, X-ray emission spectroscopy (XES), ultraviolet photoelectron spectroscopy (UPS) and X-ray photoelectron spectroscopy (XPS) are very well established techniques to probe the *electronic* structure of molecular systems, and have also been implemented in the static mode on high vapour pressure liquids [118]. Their extension into the time domain is also underway, and recent femtosecond UPS results were reported on various species in solution using a high harmonic VUV source, [202–205] while the first ps XES and RIXS studies on [$Fe^{II}(bpy)_3$] have also been reported [197, 198].

In conclusion, the prospects are bright for ultrafast X-ray absorption spectroscopy studies. While the upcoming X-ray free electron lasers offer a dramatic increase in flux and time resolution, with very exciting prospects, accessability is limited since these are single user machines. Therefore, synchrotrons will remain the main source of X-ray photons for such applications, especially that they will allow preliminary and complementary studies to those carried out at XFELs.

Acknowledgments Deep thanks to all my co-workers who have contributed to the various studies presented here, and in particular to Ch. Bressler, C. Milne, W. Gawelda, V.-T. Pham, R. M. van der Veen, T. Penfold and F. van Mourik. We also acknowledge the great collaboration with the team at the MicroXAS beamline of the Swiss Light Source (PSI, Villigen): S. Johnson, P. Beaud, D. Grolimund, C. N. Borca, G. Ingold and R. Abela. Many thanks to M. Benfatto, P. D'Angelo (Rome), I. Tavernelli (Lausanne) and A. Hauser (Geneva) for useful discussions and samples.

References

1. A.H. Zewail, J. Phys. Chem. A **104**, 5660 (2000)
2. A.H. Zewail, Angew. Chem. Int. Edit. **39**, 2587 (2000)
3. R. Henderson, Q. Rev. Biophys. **28**, 171 (1995)
4. R. Srinivasan, V.A. Lobastov, C.Y. Ruan, A.H. Zewail, Helv. Chim. Acta **86**, 1763 (2003)
5. A.H. Zewail, Annu. Rev. Phys. Chem. **57**, 65 (2006)
6. V.A. Lobastov, R. Srinivasan, A.H. Zewail, P. Natl. Acad. Sci. USA **102**, 7069 (2005)
7. S. Chen, M.T. Seidel, A.H. Zewail, Angew. Chem. Int. Edit. **45**, 5154 (2006)
8. V.A. Lobastov, J. Weissenrieder, J. Tang, A.H. Zewail, Nano Lett. **7**, 2552 (2007)
9. D. Shorokhov, A.H. Zewail, Phys. Chem. Chem.Phys. **10**, 2879 (2008)
10. M. Chergui, A.H. Zewail, Chem. Phys. Chem **10**, 28 (2009)
11. A.H. Zewail, J.M. Thomas, *4D electron microscopy : imaging in space and time* (Imperial College Press; Distributed by World Scientific Pub, London Hackensack, 2010)
12. F. Carbone, B. Barwick, O.H. Kwon, H.S. Park, J.S. Baskin, A.H. Zewail, Chem. Phys. Lett. **468**, 107 (2009)
13. F. Carbone, O.H. Kwon, A.H. Zewail, Science **325**, 181 (2009)

14. B. Barwick, D.J. Flannigan, A.H. Zewail, Nature **462**, 902 (2009)
15. F. Carbone, P. Musumeci, O.J. Luiten, C. Hebert, Chem. Phys. **392**, 1 (2012)
16. J.M. Thomas, P.A. Midgley, Chem. Phys. **385**, 1 (2011)
17. L. Guerin, E. Collet, M.B.L. Cointe, M.H. Lemee-Cailleau, H. Cailleau, M. Wulff, S. Techert, S.Y. Koshihara, J. Phys. IV **114**, 99 (2004)
18. L. Guerin, E. Collet, M.H. Lemee-Cailleau, Buron-Le Cointe, M.; Cailleau, H.; Plech, A.; Wulff, M.; Koshihara, S. Y.; Luty, T. Chem. Phys. **299**, 163 (2004)
19. F. Schotte, M.H. Lim, T.A. Jackson, A.V. Smirnov, J. Soman, J.S. Olson, G.N. Phillips, M. Wulff, P.A. Anfinrud, Science **2003**, 300 (1944)
20. E. Collet, M.H. Lemee-Cailleau, Buron-Le Cointe, M.; Cailleau, H.; Wulff, M.; Luty, T.; Koshihara, S. Y.; Meyer, M.; Toupet, L.; Rabiller, P.; Techert, S. Science **300**, 612 (2003)
21. F. Schotte, P.A. Anfinrud, G. Hummer, M. Wulff, Biophys. J. **86**, 525A (2004)
22. H. Ihee, M. Lorenc, T.K. Kim, Q.Y. Kong, M. Cammarata, J.H. Lee, S. Bratos, M. Wulff, Science **309**, 1223 (2005)
23. L. Guerin, J. Hebert, M.B.L. Cointe, S. Adachi, S. Koshihara, H. Cailleau, E. Collet, Phys Rev Lett. **105** (2010)
24. S. Bratos, F. Mirloup, R. Vuilleumier, M. Wulff, J. Chem. Phys. **116**, 10615 (2002)
25. J. Davidsson, J. Poulsen, M. Cammarata, P. Georgiou, R. Wouts, G. Katona, F. Jacobson, A. Plech, M. Wulff, G. Nyman, R. Neutze, Phys. Rev. Lett. **94**, 125509 (2005)
26. M. Cammarata, M. Lorenc, T.K. Kim, J.H. Lee, Q.Y. Kong, E. Pontecorvo, Lo Russo, M.; Schiro, G.; Cupane, A.; Wulff, M.; Ihee, H. J. Chem. Phys. **124**, 124504 (2006)
27. M. Christensen, K. Haldrup, K. Bechgaard, R. Feidenhans'l, Q.Y. Kong, M. Cammarata, Lo Russo, M.; Wulff, M.; Harrit, N.; Nielsen, M. M. J. Am. Chem. Soc. **131**, 502 (2009)
28. K. Haldrup, M. Christensen, M. Cammarata, Q.Y. Kong, M. Wulff, S.O. Mariager, K. Bechgaard, R. Feidenhans'l, N. Harrit, M.M. Nielsen, Angew. Chem. Int. Edit. **48**, 4180 (2009)
29. T.K. Kim, J.H. Lee, M. Wulff, Q.Y. Kong, H. Ihee, Chem. Phys. Chem. **2009**, 10 (1958)
30. S.L. Johnson, C.J. Milne, Trac-Trend Anal. Chem. **29**, 497 (2010)
31. R.W. Schoenlein, S. Chattopadhyay, H.H.W. Chong, T.E. Glover, P.A. Heimann, C.V. Shank, A.A. Zholents, M.S. Zolotorev, Science **287**, 2237 (2000)
32. R.W. Schoenlein, H.H.W. Chong, T.E. Glover, P.A. Heimann, W.P. Leemans, H.A. Padmore, C.V. Shank, A.A. Zholents, M.S. Zolotorev, J.S. Corlett, Cr. Acad. Sci. IV-Phys. **2**, 1373 (2001)
33. S. Khan, K. Holldack, T. Kachel, R. Mitzner, T. Quast, Phys. Rev. Lett. **97**, 074801 (2006)
34. P. Beaud, S.L. Johnson, A. Streun, R. Abela, D. Abramsohn, D. Grolimund, F. Krasniqi, T. Schmidt, V. Schlott, G. Ingold, Phys. Rev. Lett. **99**, 174801 (2007)
35. S.L. Johnson, P. Beaud, E. Vorobeva, C.J. Milne, E.D. Murray, S. Fahy, G. Ingold, Acta Crystallogr. A **66**, 157 (2010)
36. C. Bressler, C. Milne, V.T. Pham, A. ElNahhas, R.M. van der Veen, W. Gawelda, S. Johnson, P. Beaud, D. Grolimund, M. Kaiser, C.N. Borca, G. Ingold, R. Abela, M. Chergui, Science **323**, 489 (2009)
37. H.D. Wen, N. Huse, R.W. Schoenlein, A.M. Lindenberg, J Chem Phys **131**, 234505 (2009)
38. N. Huse, H. Cho, K. Hong, L. Jamula, F.M.F. de Groot, T.K. Kim, J.K. McCusker, R. Schoenlein, J. Phys. Chem. Lett. **2**, 880 (2011)
39. V.T. Pham, T.J. Penfold, R.M. van der Veen, F. Lima, A. El Nahhas, S.L. Johnson, P. Beaud, R. Abela, C. Bressler, I. Tavernelli, C.J. Milne, M. Chergui, J. Am. Chem. Soc. **133**, 12740 (2011)
40. C. Rose-Petruck, R. Jimenez, T. Guo, A. Cavalleri, C.W. Siders, F. Raksi, J.A. Squier, B.C. Walker, K.R. Wilson, C.P.J. Barty, Nature **398**, 310 (1999)
41. C.W. Siders, A. Cavalleri, K. Sokolowski-Tinten, C. Toth, T. Guo, M. Kammler, M.H. von Hoegen, K.R. Wilson, D. von der Linde, C.P.J. Barty, Science **286**, 1340 (1999)
42. A. Cavalleri, R.W. Schoenlein, Ultrafast Dyn. Processes Semiconductors **92**, 309 (2004)
43. M. Bargheer, N. Zhavoronkov, M. Woerner, T. Elsaesser, Chem. Phys. Chem. **7**, 783 (2006)
44. K.J. Gaffney, H.N. Chapman, Science **316**, 1444 (2007)
45. C.V. Schmising, M. Bargheer, M. Woerner, T. Elsaesser, Zeitschrift Fur Kristallographie **223**, 283 (2008)

46. G. Ingold, R. Abela, P. Beaud, S.L. Johnson, U. Staub, Zeitschrift Fur Kristallographie **223**, 292 (2008)
47. S.L. Johnson, P. Beaud, C.J. Milne, F.S. Krasniqi, E.S. Zijlstra, M.E. Garcia, M. Kaiser, D. Grolimund, R. Abela, G. Ingold, Phys. Rev. Lett. **100**, 155501 (2008)
48. C.V.K. Schmising, M. Bargheer, M. Kiel, N. Zhavoronkov, M. Woerner, T. Elsaesser, I. Vrejoiu, D. Hesse, M. Alexe, Phys. Rev. Lett. **98**, 248301 (2007)
49. M. Woerner, F. Zamponi, Z. Ansari, J. Dreyer, B. Freyer, M. Premont-Schwarz, T. Elsaesser, J. Chem. Phys. **133**, 064509 (2010)
50. F. Zamponi, Z. Ansari, M. Woerner, T. Elsaesser, Opt. Express **18**, 947 (2010)
51. M. BenNun, J.S. Cao, K.R. Wilson, J. Phys. Chem. A **101**, 8743 (1997)
52. J.S. Cao, K.R. Wilson, J. Phys. Chem. A **102**, 9523 (1998)
53. J.P. Bergsma, M.H. Coladonato, P.M. Edelsten, J.D. Kahn, K.R. Wilson, D.R. Fredkin, J. Chem. Phys. **84**, 6151 (1986)
54. H. Ihee, Accounts. Chem. Res. **42**, 356 (2009)
55. L.X. Chen, J. Electron Spectrosc. **119**, 161 (2001)
56. L.X. Chen, Angew. Chem. Int. Edit. **43**, 2886 (2004)
57. C. Bressler, M. Chergui, Chem. Rev. **104**, 1781 (2004)
58. C. Bressler, R. Abela, M. Chergui, Zeitschrift Fur Kristallographie **223**, 307 (2008)
59. L.X. Chen, X.Y. Zhang, J.V. Lockard, A.B. Stickrath, K. Attenkofer, G. Jennings, D.J. Liu, Acta Crystallogr. A **66**, 240 (2010)
60. C. Bressler, M. Chergui, Annu. Rev. Phys. Chem. **61**, 263 (2010)
61. M. Chergui, Acta Crystallogr. A **66**, 229 (2010)
62. D.C. Koningsberger, R. Prins, *X-ray absorption : principles, applications, techniques of EXAFS, SEXAFS, and XANES* (Wiley, New York, 1988)
63. J. Stöhr, *NEXAFS spectroscopy* (Springer, Berlin etc., 1992)
64. J.J. Rehr, R.C. Albers, Rev. Mod. Phys. **72**, 621 (2000)
65. E.A. Stern, Phys. Rev. B **10**, 3027 (1974)
66. D.E. Sayers, E.A. Stern, F.W. Lytle, Phys. Rev. Lett. **27**, 1204 (1971)
67. A. Filipponi, J. Phys.-Condens. Mat. **6**, 8415 (1994)
68. A.L. Ankudinov, C.E. Bouldin, J.J. Rehr, J. Sims, H. Hung, Phys. Rev. B **65**, 104 (2002)
69. J.J. Rehr, J.J. Kas, F.D. Vila, M.P. Prange, K. Jorissen, Phys. Chem. Chem. Phys. **12**, 5503 (2010)
70. A.L. Ankudinov, B. Ravel, J.J. Rehr, S.D. Conradson, Phys. Rev. B **58**, 7565 (1998)
71. C.R. Natoli, M. Benfatto, Della Longa, S.; Hatada, K. J. Synchrotron Radiat. **10**, 26 (2003)
72. J.J. Rehr, J.J. Kas, M.P. Prange, A.P. Sorini, Y. Takimoto, F. Vila, C.R. Phys. **10**, 548 (2009)
73. B. Ravel, M. Newville, J. Synchrotron Radiat. **12**, 537 (2005)
74. J.L. Beeby, Proc. R Soc. Lon. Ser.-A **279**, 82 (1964)
75. C.A. Ashley, S. Doniach, Phys. Rev. B **11**, 1279 (1975)
76. P. Lloyd, P.V. Smith, Adv. Phys. **21**, 69 (1972)
77. T.A. Tyson, M. Benfatto, C.R. Natoli, B. Hedman, K.O. Hodgson, Phys. B **158**, 425 (1989)
78. T.A. Tyson, K.O. Hodgson, C.R. Natoli, M. Benfatto, Phys. Rev. B **46**, 5997 (1992)
79. C.R. Natoli, M. Benfatto, S. Doniach, Phys. Rev. A **34**, 4682 (1986)
80. M. Benfatto, Della Longa, S.; Hatada, K.; Hayakawa, K.; Gawelda, W.; Bressler, C.; Chergui, M. J. Phys. Chem. B **110**, 14035 (2006)
81. S.I. Zabinsky, J.J. Rehr, A. Ankudinov, R.C. Albers, M.J. Eller, Phys. Rev. B **52**, 2995 (1995)
82. J. Mustre, Y. Yacoby, E.A. Stern, J.J. Rehr, Phys. Rev. B **42**, 10843 (1990)
83. W. Gawelda, V.T. Pham, R.M. van der Veen, D. Grolimund, R. Abela, M. Chergui, C. Bressler, J. Chem. Phys. **130**, 124520 (2009)
84. K. Okada, A. Kotani, B.T. Thole, J. Electron Spectrosc. **58**, 325 (1992)
85. B.T. Thole, G. Vanderlaan, J.C. Fuggle, G.A. Sawatzky, R.C. Karnatak, J.M. Esteva, Phys. Rev. B **32**, 5107 (1985)
86. F. De Groot, Coordin. Chem. Rev. **249**, 31 (2005)

87. F. De Groot, A. Kotani, *Core Level Spectroscopy of Solids* (Taylor & Francis, New York, 2008)
88. W. Gawelda, M. Johnson, F.M.F. de Groot, R. Abela, C. Bressler, M. Chergui, J. Am. Chem. Soc. **128**, 5001 (2006)
89. F. De Groot, Chem. Rev. **101**, 1779 (2001)
90. A. Bianconi, In X-ray absorption principles, applications, techniques of exafs, sexafs and xanes, ed. by D. C. Koningsberger, R. Prins, (Wiley, New York, 1988), p 12
91. M. Benfatto, Della Longa, S.; Natoli, C. R. J. Synchrotron Radiat. **10**, 51 (2003)
92. M. Benfatto, D.S. Longa, J. Synchrotron Radiat. **1087**, 8 (2001)
93. W. Gawelda, V.T. Pham, A. El Nahhas, M. Kaiser, Y. Zaushytsin, S. Johnson, D. Grolimund, R. Abela, A. Hauser, C. Bressler, M. Chergui, AIP Conf. Proc. **882**, 31 (2007)
94. A.A. Zholents, M.S. Zolotorev, Phys. Rev. Lett. **76**, 912 (1996)
95. K. Murakami, H.C. Gerritsen, H. Vanbrug, F. Bijkerk, F.W. Saris, M.J. Vanderwiel, Phys. Rev. Lett. **56**, 655 (1986)
96. K. O. T. Murakami, A. Miyashita, O. Yoda, AIP Conference Proceedings, Vol. 228, p. 375 (1994)
97. F. Raksi, K.R. Wilson, Z.M. Jiang, A. Ikhlef, C.Y. Cote, J.C. Kieffer, J. Chem. Phys. **104**, 6066 (1996)
98. I.V. Tomov, D.A. Oulianov, P.L. Chen, P.M. Rentzepis, J. Phys. Chem. B **103**, 7081 (1999)
99. F. Benesch, T.W. Lee, Y. Jiang, C.G. Rose-Petruck, Opt. Lett. **2004**, 29 (1028)
100. K. Oguri, Y. Okano, T. Nishikawa, H. Nakano, Appl. Phys. Lett. **87**, 011503 (2005)
101. Y. Okano, K. Oguri, T. Nishikawa, H. Nakano, Rev. Sci. Instrum. **77**, 046105 (2006)
102. K. Hatanaka, T. Miura, H. Fukumura, Appl. Phys. Lett. **80**, 3925 (2002)
103. T. Lee, Y. Jiang, C.G. Rose-Petruck, F. Benesch, J. Chem. Phys. **122**, 84506 (2005)
104. T. Guo, Laser Photonics Rev. **3**, 591 (2009)
105. T. Elsaesser, M. Woerner, Acta Crystallogr. A **66**, 168 (2010)
106. C. Bressler, M. Saes, M. Chergui, D. Grolimund, R. Abela, P. Pattison, J. Chem. Phys. **116**, 2955 (2002)
107. C. Bressler, M. Saes, M. Chergui, R. Abela, P. Pattison, Nucl. Instrum. Meth. A **467**, 1444 (2001)
108. M. Saes, W. Gawelda, M. Kaiser, A. Tarnovsky, C. Bressler, M. Chergui, S.L. Johnson, D. Grolimund, R. Abela, Synchrotron Radiat. News **16**, 12 (2003)
109. M. Saes, C. Bressler, F. van Mourik, W. Gawelda, M. Kaiser, M. Chergui, C. Bressler, D. Grolimund, R. Abela, T.E. Glover, P.A. Heimann, R.W. Schoenlein, S.L. Johnson, A.M. Lindenberg, R.W. Falcone, Rev. Sci. Instrum. **75**, 24 (2004)
110. W. Gawelda, C. Bressler, M. Saes, M. Kaiser, A. Tarnovsky, D. Grolimund, S.L. Johnson, R. Abela, M. Chergui, Phys. Scr. **T115**, 102 (2005)
111. T. Giessel, D. Brocker, P. Schmidt, W. Widdra, Rev. Sci. Instrum. **74**, 4620 (2003)
112. W. Widdra, D. Brocker, T. Giessel, I.V. Hertel, W. Kruger, A. Liero, F. Noack, V. Petrov, D. Pop, P.M. Schmidt, R. Weber, I. Will, B. Winter, Surf. Sci. **543**, 87 (2003)
113. P. Fons, D. Brewe, E. Stern, A.V. Kolobov, J. Tominaga, Mater Res. Soc. Symp. P **918**, 147 (2006)
114. E.A. Stern, D. Brewe, AIP Conf. Proc. **882**, 24 (2007)
115. F.A. Lima, C.J. Milne, D.C.V. Amarasinghe, M.H. Rittmann-Frank, R.M. van der Veen, M. Reinhard, V.T. Pham, S. Karlsson, S.L. Johnson, D. Grolimund, C. Borca, T. Huthwelker, M. Janousch, F. van Mourik, R. Abela, M. Chergui, Rev. Sci. Instrum. **82**, 063111 (2011)
116. F.M.F. De Groot, J. Vogel, (Oxford University Press, Oxford, 2004)
117. A. Nilsson, L.G.M. Pettersson, Chem. Phys. **389**, 1 (2011)
118. B. Winter, M. Faubel, Chem. Rev. **106**, 1176 (2006)
119. G. Gavrila, K. Godehusen, C. Weniger, E.T.J. Nibbering, T. Elsaesser, W. Eberhardt, P. Wernet, Appl. Phys. A-Mater. **96**, 11 (2009)
120. N. Huse, Private communication
121. D.M. Mills, A. Lewis, A. Harootunian, J. Huang, B. Smith, Science **223**, 811 (1984)

122. H.B. Gray, J.R. Winkler, Annu. Rev. Biochem. **65**, 537 (1996)
123. H. B. Gray, *In Electron transfer in chemistry,* ed. by V. Balzani, (Wiley-VCH, Weinheim, 2001)
124. A. Juris, V. Balzani, F. Barigelletti, S. Campagna, P. Belser, A. Vonzelewsky, Coordin. Chem. Rev. **84**, 85 (1988)
125. M. Gratzel, Nature **414**, 338 (2001)
126. M. Saes, C. Bressler, R. Abela, D. Grolimund, S.L. Johnson, P.A. Heimann, M. Chergui, Phys. Rev. Lett. **90**, 047403 (2003)
127. A.N. Tarnovsky, W. Gawelda, M. Johnson, C. Bressler, M. Chergui, J. Phys. Chem. B **110**, 26497 (2006)
128. F.M.F. De Groot, Z.W. Hu, M.F. Lopez, G. Kaindl, F. Guillot, M. Tronc, J. Chem. Phys. **101**, 6570 (1994)
129. G. Calzaferri, R. Rytz, J. Phys. Chem.-Us **99**, 12141 (1995)
130. E. König, K.J. Watson, Chem. Phys. Lett. **6**, 457 (1970)
131. K. Nozaki, K. Takamori, Y. Nakatsugawa, T. Ohno, Inorg. Chem. **45**, 6161 (2006)
132. F. Alary, J.L. Heully, L. Bijeire, P. Vicendo, Inorg. Chem. **46**, 3154 (2007)
133. G.B. Shaw, C.D. Grant, H. Shirota, E.W. Castner, G.J. Meyer, L.X. Chen, J. Am. Chem. Soc. **129**, 2147 (2007)
134. M. Iwamura, S. Takeuchi, T. Tahara, J. Am. Chem. Soc. **129**, 5248 (2007)
135. M. Iwamura, H. Watanabe, K. Ishii, S. Takeuchi, T. Tahara, Springer Ser. Chem. **92**, 382 (2009)
136. L.X. Chen, G.B. Shaw, I. Novozhilova, T. Liu, G. Jennings, K. Attenkofer, G.J. Meyer, P. Coppens, J. Am. Chem. Soc. **125**, 7022 (2003)
137. G. Smolentsev, A.V. Soldatov, L.X. Chen, J. Phys. Chem. A **112**, 5363 (2008)
138. T.J. Penfold, S. Karlsson, G. Capano, F.A. Lima, J. Rittmann, M.H. Rittmann-Frank, M. Reinhard, O. Bräm, E. Baranoff, R. Abela, I. Tavernelli, U. Röthlisberger, C. Milne, M. Chergui, J. Phys. Chem. A 117, 4591–4601 (2013)
139. T. J. Penfold, B.F.E. Curchod, I. Tavernelli, R. Abela, U. Röthlisberger, M. Chergui, Phys. Chem. Chem. Phys. 14, 9444–9450 (2012)
140. A. Vlcek, Coordin. Chem. Rev. **200**, 933 (2000)
141. R.M. van der Veen, C.J. Milne, A. El Nahhas, F.A. Lima, V.T. Pham, J. Best, J.A. Weinstein, C.N. Borca, R. Abela, C. Bressler, M. Chergui, Angew. Chem. Int. Edit. **48**, 2711 (2009)
142. I.V. Novozhilova, A.V. Volkov, P. Coppens, J. Am. Chem. Soc. **2003**, 125 (1079)
143. J.V. Lockard, A.A. Rachford, G. Smolentsev, A.B. Stickrath, X.H. Wang, X.Y. Zhang, K. Atenkoffer, G. Jennings, A. Soldatov, A.L. Rheingold, F.N. Castellano, L.X. Chen, J. Phys. Chem. A **114**, 12780 (2010)
144. R.M. van der Veen, C. Bressler, C.J. Milne, V.T. Pham, A. El Nahhas, F.A. Lima, W. Gawelda, C.N. Borca, R. Abela, M. Chergui, J. Phys. Conf. Ser. **190**, 012052 (2009)
145. A. Hauser, *Spin Crossover in Transition Metal Compounds II.*(Springer, Berlin, 2004), Vol. 234, p. 155
146. B. Ordejon, C. de Graaf, C. Sousa, J. Am. Chem. Soc. **130**, 13961 (2008)
147. P. Guionneau, M. Marchivie, G. Bravic, J.F. Letard, D. Chasseau, Top. Curr. Chem. **234**, 97 (2004)
148. H. Oyanagi, T. Tayagaki, K. Tanaka, J. Phys. Chem. Solids **65**, 1485 (2004)
149. H. Oyanagi, T. Tayagaki, K. Tanaka, J. Lumin. **119**, 361 (2006)
150. L.M.L. Daku, A. Vargas, A. Hauser, A. Fouqueau, M.E. Casida, Chem. Phys. Chem. **6**, 1393 (2005)
151. M. Khalil, M.A. Marcus, A.L. Smeigh, J.K. McCusker, H.H.W. Chong, R.W. Schoenlein, J. Phys. Chem. A **110**, 38 (2006)
152. W. Gawelda, V.T. Pham, M. Benfatto, Y. Zaushytsin, M. Kaiser, D. Grolimund, S. Johnson, R. Abela, C. Bressler, M. Chergui, Phys. Rev. Lett. **98**, 057401 (2007)
153. J.E. Monat, J.K. McCusker, J. Am. Chem. Soc. **122**, 4092 (2000)
154. E.A. Juban, A.L. Smeigh, J.E. Monat, J.K. McCusker, Coordin. Chem. Rev. **250**, 1783 (2006)
155. J.K. McCusker, A.L. Rheingold, D.N. Hendrickson, Inorg. Chem. **35**, 2100 (1996)
156. W. Gawelda, A. Cannizzo, V.T. Pham, F. van Mourik, C. Bressler, M. Chergui, J. Am. Chem. Soc. **129**, 8199 (2007)

157. V. Briois, P. Sainctavit, G.J. Long, F. Grandjean, Inorg. Chem. **40**, 912 (2001)
158. J.P. Tuchagues, A. Bousseksou, G. Molnar, J.J. McGarvey, F. Varret, Top. Curr. Chem. **235**, 85 (2004)
159. A. Cannizzo, C.J. Milne, C. Consani, W. Gawelda, C. Bressler, F. van Mourik, M. Chergui, Coordin. Chem. Rev. **254**, 2677 (2010)
160. O. Bram, A. Cannizzo, A.A. Oskouei, A. Tortschanoff, F. van Mourik, M. Chergui, Springer Ser. Chem. **92**, 346 (2009)
161. N. Huse, H.D. Wen, D. Nordlund, E. Szilagyi, D. Daranciang, T.A. Miller, A. Nilsson, R.W. Schoenlein, A.M. Lindenberg, Phys. Chem. Chem. Phys. **11**, 3951 (2009)
162. P. Wernet, G. Gavrila, K. Godehusen, C. Weniger, E.T.J. Nibbering, T. Elsaesser, W. Eberhardt, Appl. Phys. A-Mater **92**, 511 (2008)
163. M. Maroncelli, J. Mol. Liq. **57**, 1 (1993)
164. R.M. Stratt, M. Maroncelli, J. Phys. Chem.-Us **100**, 12981 (1996)
165. S.K. Pal, A.H. Zewail, Chem. Rev. **104**, 2099 (2004)
166. V.T. Pham, W. Gawelda, Y. Zaushytsin, M. Kaiser, D. Grolimund, S. Johnson, R. Abela, C. Bressler, M. Chergui, J. Am. Chem. Soc. **129**, 1530 (2007)
167. C.G. Elles, I.A. Shkrob, R.A. Crowell, D.A. Arms, E.C. Landahl, J. Chem. Phys. **128**, 061102 (2008)
168. V.T. Pham, I. Tavernelli, C.J. Milne, R.M. van der Veen, P. D'Angelo, C. Bressler, M. Chergui, Chem. Phys. **371**, 24 (2010)
169. K.M. Kadish, K.M. Smith, R. Guilard, *The porphyrin handbook* (Academic Press, San Diego, 2000)
170. L.X. Chen, P.L. Lee, D. Gosztola, W.A. Svec, P.A. Montano, M.R. Wasielewski, J. Phys. Chem. B **103**, 3270 (1999)
171. L.X. Chen, W.J.H. Jager, G. Jennings, D.J. Gosztola, A. Munkholm, J.P. Hessler, Science **292**, 262 (2001)
172. L.X. Chen, X.Y. Zhang, E.C. Wasinger, K. Attenkofer, G. Jennings, A.Z. Muresan, J.S. Lindsey, J. Am. Chem. Soc. **129**, 9616 (2007)
173. S. Della-Longa, L.X. Chen, P. Frank, K. Hayakawa, K. Hatada, M. Benfatto, Inorg. Chem. **48**, 3934 (2009)
174. D.M. Mills, Acta Crystallogr. A **40**, C393 (1984)
175. A. Clozza, A.C. Castellano, S. Dellalonga, A. Giovannelli, A. Bianconi, Rev. Sci. Instrum. **60**, 2519 (1989)
176. M.R. Chance, M.D. Wirt, E.M. Scheuring, L.M. Miller, A.H. Xie, D.E. Sidelinger, Rev. Sci. Instrum. **64**, 2035 (1993)
177. L. Powers, B. Chance, M. Chance, B. Campbell, J. Friedman, S. Khalid, C. Kumar, A. Naqui, K.S. Reddy, Y. Zhou, Biochem.-Us **26**, 4785 (1987)
178. A. Arcovito, D.C. Lamb, G.U. Nienhaus, J.L. Hazemann, M. Benfatto, S.D. Longa, Biophys. J. **88**, 2954 (2005)
179. H.X. Wang, G. Peng, S.P. Cramer, J. Electron Spectrosc. **143**, 1 (2005)
180. J.L. Martin, M.H. Vos, Annu. Rev. Biophys. Biomol. Struct. **21**, 199 (1992)
181. M. Vos, Actual Chimique, **52** (2008)
182. M.H. Vos, Bba-Bioenerg. **1777**, 15 (2008)
183. M. Chergui, *Advances in Accounting Behavioral Research (BIPH)* (Elsevier: Amsterdam, 2011) (Vol. Comprehensive Biophysics)
184. F. A. Lima, E. A. In preparation (2012)
185. A. Cavalleri, H.H.W. Chong, S. Fourmaux, T.E. Glover, P.A. Heimann, J.C. Kieffer, B.S. Mun, H.A. Padmore, R.W. Schoenlein, Phys. Rev. B **69**, 153106 (2004)
186. A. Cavalleri, M. Rini, H.H.W. Chong, S. Fourmaux, T.E. Glover, P.A. Heimann, J.C. Kieffer, R.W. Schoenlein, Phys. Rev. Lett. **95**, 067405 (2005)
187. A. Cavalleri, S. Wall, C. Simpson, E. Statz, D.W. Ward, K.A. Nelson, M. Rini, R.W. Schoenlein, Nature **442**, 664 (2006)
188. B.W. Adams, M.F. DeCamp, E.M. Dufresne, D.A. Reis, Rev. Sci. Instrum. **73**, 4150 (2002)

189. S.L. Johnson, P.A. Heimann, A.M. Lindenberg, H.O. Jeschke, M.E. Garcia, Z. Chang, R.W. Lee, J.J. Rehr, R.W. Falcone, Phys. Rev. Lett. **91**, 157403 (2003)

190. S.L. Johnson, P.A. Heimann, A.G. MacPhee, A.M. Lindenberg, O.R. Monteiro, Z. Chang, R.W. Lee, R.W. Falcone, Phys. Rev. Lett. **94**, 057407 (2005)

191. K. Oguri, H. Nakano, T. Nishikawa, N. Uesugi, Appl. Phys. Lett. **79**, 4506 (2001)

192. Y. Okano, K. Oguri, T. Nishikawa, H. Nakano, J. Appl. Phys. **99**, 063302 (2006)

193. E. Vorobeva, S.L. Johnson, P. Beaud, C.J. Milne, M. Benfatto, G. Ingold, Phys. Rev. B **80**, 134301 (2009)

194. C. Stamm, T. Kachel, N. Pontius, R. Mitzner, T. Quast, K. Holldack, S. Khan, C. Lupulescu, E.F. Aziz, M. Wietstruk, H.A. Durr, W. Eberhardt, Nat. Mater. **6**, 740 (2007)

195. M. Haumann, P. Liebisch, C. Muller, M. Barra, M. Grabolle, H. Dau, Science **2005**, 310 (1019)

196. M. Haumann, C. Muller, P. Liebisch, T. Neisius, H. Dau, J. Synchrotron Radiat. **12**, 35 (2005)

197. G. Vanko, P. Glatzel, V.T. Pham, R. Abela, D. Grolimund, C.N. Borca, S.L. Johnson, C.J. Milne, C. Bressler, Angew. Chem. Int. Edit. **49**, 5910 (2010)

198. C. Bressler, Private communication

199. C. Bressler, R. Abela, M. Cammarata, Private communication

200. E.F. Aziz, N. Ottosson, S. Bonhommeau, N. Bergmann, W. Eberhardt, M. Chergui, Phys. Rev. Lett. **102**, 68103 (2009)

201. E.F. Aziz, M.H. Rittmann-Frank, K.M. Lange, S. Bonhommeau, M. Chergui, Nat. Chem. **2**, 853 (2010)

202. O. Link, E. Lugovoy, K. Siefermann, Y. Liu, M. Faubel, B. Abel, Appl. Phys. A-Mater. **96**, 117 (2009)

203. A. Lubcke, F. Buchner, N. Heine, I.V. Hertel, T. Schultz, Phys. Chem. Chem. Phys. **12**, 14629 (2010)

204. Y. Tang, H. Shen, K. Sekiguchi, N. Kurahashi, T. Mizuno, Y.I. Suzuki, T. Suzuki, Phys. Chem. Chem. Phys. **12**, 3653 (2010)

205. Y. Tang, Y.I. Suzuki, H. Shen, K. Sekiguchi, N. Kurahashi, K. Nishizawa, P. Zuo, T. Suzuki, Chem. Phys. Lett. **494**, 111 (2010)

206. R.M. van der Veen, C.J. Milne, V.T. Pham, A. El Nahhas, J.A. Weinstein, J. Best, C.N. Borca, C. Bressler, M. Chergui, Chimia **62**, 287 (2008)

207. R.M. van der Veen, J.J. Kas, C.J. Milne, V.T. Pham, A. El Nahhas, F.A. Lima, D.A. Vithanage, J.J. Rehr, R. Abela, M. Chergui, Phys. Chem. Chem. Phys. **12**, 5551 (2010)

208. S.J. Lind, K.C. Gordon, M.R. Waterland, J. Raman Spectrosc. **39**, 1556 (2008)

209. L.X. Chen, G. Jennings, T. Liu, D.J. Gosztola, J.P. Hessler, D.V. Scaltrito, G.J. Meyer, J. Am. Chem. Soc. **124**, 10861 (2002)

Chapter 2
In-situ X-ray Diffraction at Synchrotrons and Free-Electron Laser Sources

Vedran Vonk and Heinz Graafsma

Abstract X-ray Diffraction (XRD) is an outstanding tool for structural analyses at the atomic scale, and both the experimental techniques and the theoretical interpretations are well established. X-rays also have the advantage of being highly penetrating, as compared to electrons for instance, allowing for the study of bulk materials, or to study samples in complicated environments. The high photon fluxes available at third generation synchrotron sources make it possible to collect full diffraction patterns in relatively short times, and thus to follow time varying processes in-situ. In the first part of this chapter we briefly discuss the advantages and disadvantages of X-rays as compared to other probes like electrons or neutrons. In the second part as an example in-situ surface X-ray diffraction studies of growing films using pulsed laser deposition (PLD) will be presented. The hetero-epitaxial growth process, especially of the first mono-layers can only be understood by in-situ diffraction studies in the PLD chamber under deposition conditions. Also high energy diffraction of buried interfaces will be discussed briefly. The final part of this chapter will present the possibilities for in-situ diffraction studies at the upcoming Free-Electron Laser sources, with fully coherent beams and sufficient intensities to collect full diffraction patterns with

V. Vonk (✉)
Institute for Molecules and Materials, Radboud University Nijmegen, Nijmegen, The Netherlands
e-mail: vedran.vonk@desy.de

V. Vonk
Photon-Science Nanolaboratory, Deutches Elektronen-Synchrotron, DESY, Hamburg, Germany

H. Graafsma
Photon-Science Detector Group, Deutches Elektronen-Synchrotron, DESY, Notkestr. 85, D-22607 Hamburg, Germany
e-mail: heinz.graafsma@desy.de

H. Graafsma
STC Research Center, University of Mid-Sweden, 85107 Sundsvall, Sweden

A. Ziegler et al. (eds.), *In-situ Materials Characterization*,
Springer Series in Materials Science 193, DOI: 10.1007/978-3-642-45152-2_2,
© Springer-Verlag Berlin Heidelberg 2014

single 100 femto-second pulses. The characteristics of the Free-Electron-Lasers and various planned experiments will be presented.

2.1 The Need for In-situ X-ray Observations

Many phenomena related to the solid state require a description at the atomic level, since this is the smallest relevant and therefore most fundamental length scale. Of particular interest are physical processes, such as phase transitions and crystal growth. These lie at the basis of many important applications in the chemical industry and (electrical) engineering. Motivations for obtaining atomic-scale insight include process optimization and the development of new techniques and materials by gaining fundamental knowledge. Often, however, there is a lack of such fundamental knowledge, and new experiments and theories need to be devised. There is a trend that both the time and the length scale on which the solid state is understood are getting smaller. The former development aims at picturing different steps of atomic interactions, whereas the latter uncovers the influence of surfaces and interfaces on physical properties. Both trends are experimentally challenging in the sense that decreasing exposure times and decreasing sample volumes reduce signal levels tremendously. Here lies an opportunity for third generation synchrotron sources and X-ray Free Electron Lasers.

Many of the open questions relate to the atomic order and how this is influenced by external factors. An example of this is crystallization, where a particular atom or molecule transforms from being completely disordered in its solution or gas phase to a crystalline form. A more subtle example is the polarization switching in ferro-electric materials, where one atom may occupy two different positions which determine the direction of polarization, and where temperature-induced hopping over these two sites may result in a net zero polarization. In the past decades, to both examples many experimental studies are devoted. Such studies have extracted spatial- and time-averaged information from diffraction experiments, and are therefore representative for the stationary bulk behavior of the material.

In order to follow what is happening as a function of the thermodynamic parameters, one needs a probe that does not interfere too much with the process itself, and that is able to reach and leave the area of interest. In many cases, the probe consists of charged particles. However, deeply buried structures and/or high gas pressures do not always allow for such experiments. In those cases, only X-rays and neutrons would qualify. Since the latter probe lacks the availability of high flux sources, only the former remains. The most brilliant X-ray sources are synchrotrons and X-ray free electron lasers.

Many of the solid-state theories that have been developed so far, describe properties in the bulk, thereby ignoring the effects of microscopic defects and surfaces. The former of these have been studied intensively for the past 50 years using electron microscopy, which has provided many important insights. The latter is at the heart of surface science, where chemistry and physics meet in order to obtain detailed information on the microscopic nature. In both cases, in-situ studies are severely hindered, either by the destructive nature or the requirements of vacuum. X-rays could in principle be used for both types of studies. By special optics in combination with adapted geometries, it is possible to measure the scattering coming from a volume of roughly $1 \times 1 \times 1 \ \mu m^3$ [1]. This volume is getting ever smaller, and it is to be expected that in the future it could become as small as $10 \times 10 \times 10 \ nm^3$. Such a small probe is still not comparable with electron beams, which can be focused down to the Angstrom scale, but might fill in an important gap.

The physics that can be addressed with small X-ray beams is that of condensed matter in small ensembles. This might deviate a lot from the description of average, bulk properties, which are often well understood by equilibrium thermodynamics.

Buried interfaces appear in a large variety in nature. One can think of the solid–solid ice–rock interface, of which the properties are important for avalanche science. The solid–liquid rock–water interface is omnipresent on our planet, and in particular the chemistry that occurs when the water pressure amounts up to several hundreds of atmospheres, as appears on the bottom of oceans. It is known that chemical reactions proceed completely differently under these circumstances, because of the changed acidity of water at these pressures. Therefore, the study of buried interfaces is not only of fundamental interest, but also highly relevant to practical and applied sciences.

2.2 Advantages and Disadvantages of In-situ X-ray Diffraction

As already mentioned in the previous paragraph there are a certain number of advantages of using X-rays as probe rather than, for instances, electrons. These advantages originate from the relatively low scattering cross-section of X-ray photons, as indicated in Fig. 2.1.

The first consequence of this low scattering cross-section is that the Born approximation can be applied. This means that the scattered wave field by a particle is negligibly small compared to the incident wave field on that particle. As a consequence, one can take the incident field instead of the total field as the driving force for all particles (atoms) in the sample, thereby greatly simplifying the theoretical interpretation of the scattering pattern. This so called kinematical diffraction theory is well developed and applicable to nearly all practical samples. The only exceptions are highly perfect single crystals like silicon or germanium. Electrons, on the other hand, have much higher scattering cross-sections as seen from Fig. 2.1. This means that the scattered field by an atom is not any longer negligible to the incident wave field, and consequently, one is obliged to take the total field rather than simply the incident field at each scattering atom. In other words, in order to calculate the wave field at one atom, one has to take into account the incident field, as well as the scattered field from all other atoms in the sample. This is known as dynamical scattering. It is clear that this makes the interpretation of the scattered intensities extremely complicated, if not impossible.

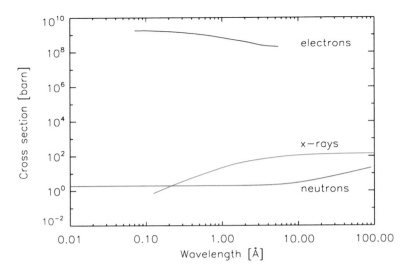

Fig. 2.1 Calculated scattering cross sections in forward direction for silicon [38]. Shown are the elastic, coherent cross sections for electrons (*blue*), X-rays (*green*) and neutrons (*red*) given in barn ($=10^{-28}$ m^2) as a function of the wavelength of the particular probe. The wavelength regime relevant to diffraction, i.e. around 1 Å is shown. The scattering interaction of electrons is orders of magnitude larger because of the strong coulomb interaction, which is much weaker for X-rays and even absent for neutrons

Another advantage of the low cross section of X-rays is the much larger penetration depth as compared to electrons. Electrons are fully absorbed by a few centimeters of atmospheric air, which means that measurements have to be performed under ultra-high or high vacuum conditions. High energy, >15 keV, X-rays, on the other hand have negligible absorption in air. For 22 keV X-rays, even a sample chamber of 0.5 mm aluminum, will only absorbed 50 % of the photons. This means that in-situ experiments under real life conditions in sophisticated sample chambers are possible. As indicated in the two examples later in this chapter, this is often the only way to get detailed information of important processes.

The negative aspect of the low scattering cross-section is the inefficient use of available photons. X-rays that pass through the sample without interaction will not contribute to the signal. Therefore, in order to get a statistically significant signal in a short time, one has to use powerful X-ray sources. Fortunately they are available at Storage Ring and Free-Electron Laser Synchrotron sources. Unfortunately this means that these experiments can no longer be build-up and performed at the home laboratories, which makes access, especially for students, more complicated. Access to these large facilities is regulated via a peer review proposal system, and turn-around times between submitting proposals and doing experiments count in months to 1 year, rather than days. Also the instrumentation involved with in-situ X-ray scattering sample chambers and experiments should not be underestimated. Often long term involvement and investments are needed before the required experimental set-up is fully commissioned and routinely available. This means that in-situ Synchrotron X-ray scattering experiments are generally reserved for systems of major importance, which cannot be studied by laboratory based techniques.

2.3 Selected Examples of In-situ X-ray Diffraction

In this section we will present in detail an X-ray surface diffraction study of growing thin films inside a Pulsed-Laser Deposition (PLD) chamber at a synchrotron storage ring. We will also present some results on high-energy micro-diffraction studies of deeply buried interfaces. It is understood that these are just two of many possible examples, but should provide the reader a reasonable overview of the possibilities.

2.3.1 Complex Transition Metal Oxides

Complex transition metal oxides are an important class of materials due to the wealth of physical properties they comprise. It is already known for several decades that these compounds can show a multitude of technologically interesting properties, such as ferromagnetism, magnetoresistance, ferroelectricity and superconductivity. Often, the phase-diagram of such a compound shows several of the aforementioned properties, whereby phase-transitions occur as one changes the doping level of one of the cations or oxygens. The changes in stoichiometry can be minute, but the properties may change drastically. From an engineering point of view, planar structures are often desired, which has resulted in an ever growing effort to make thin film structures [2]. Bulk materials are mostly produced by high temperature sintering, resulting in polycrystalline materials. In thin film form, however, these materials can be grown with a high degree of perfection.

Although several thin film growth techniques exist, Pulsed Laser Deposition (PLD) has proved the most successful for complex oxides over the last decades. This is in part because of its unique ability for operation in the required temperature-oxygen partial pressure regime, but also because of its pulsed nature, which allows for a precise thickness control. At present, thin films of all kinds of complex oxides are routinely grown throughout the world. The fact that the thickness of these films can be controlled at the unit cell level, has resulted in two important abilities of thin film oxides. First of all, with PLD being a kinetic growth technique, it is possible to engineer artificial structures. These consist for example of multilayers of different materials, which can not be made with growth techniques that rely on thermodynamic equilibria. Such structures enable the combination or even the emergence of new physical properties. A second important feature of thin film complex oxides at the nanometer-scale is that the presence of surfaces, interfaces and strain adds new dimensions to their phase-diagrams. The subject of nano-scale engineering has become one of the major aspects of complex oxide research. It focuses on the fundamental understanding of the role of surfaces and interfaces for physical properties, thereby deepening the knowledge of the atomic structure in relation to these properties.

2.3.1.1 Pulsed Laser Deposition (PLD)

PLD is a thin film growth technique, which makes use of a powerful laser to ablate target material, which is subsequently deposited on a heated single crystal surface. The ablation process produces a plasma composing of atomic clusters, a part of which are charged. The physical processes occurring during the interaction of the laser with the target are not fully understood yet. During the arrival of the plasma at the substrate surface, the driving force for crystallization, i.e. the supersaturation, is very large because the state of the matter in the plasma is very far from the equilibrium condensed matter state. As a consequence, the crystal growth is to a very large extent governed by kinetics. Compared to molecular beam epitaxy, the deposition rate during PLD is extremely high of the order of 10^5 monolayers/s, because the actual crystallization occurs only in a very short time interval of the order of microseconds.

The growth dynamics during PLD have been extensively studied using Reflection High Energy Electron Diffraction (RHEED) [3], even at high background pressure [4]. Due to the strong interaction of the electrons with the crystal, mainly information about the surface morphology can be extracted with RHEED. Structural information of the grown films is then obtained ex-situ in a sort of archaeological way, and often not at deposition conditions, which may have a large influence on the atomic and micro structure. Obtaining information about the atomic and micro structure during growth requires another technique and for this purpose in-situ Surface X-ray Diffraction is suited. By the use of X-rays, the quantitative interpretation of the scattering patterns is much facilitated compared with electrons, because the so-called kinematical scattering approximation is valid. The less interacting X-rays penetrate more into the material, enabling to follow the structure evolution of the complete film during growth. At the same time the surface roughness can be probed, which allows to follow the dynamics. Up to now, several in-situ surface sensitive X-ray diffraction measurements during PLD were reported [5–7]. These have concentrated mostly on the growth dynamics. The fact that the scattering signals could be modeled using kinematical theory, enabled the disentanglement of structure and morphology. The most important findings are that just after the arrival of the plasma plume at the substrate, there exists a superfast relaxation mechanism [6], with a hitherto unresolved time-constant. Then there is a slower thermal kinetic smoothing mechanism, of which the characteristic time depends on the island coverage. It has been suggested that the impinging species could be energetic enough to break up two-dimensional islands, a process which is easier for smaller islands and therefore changes during the course of layer-by-layer growth.

2.3.1.2 Growth of $YBa_2Cu_3O_{7-x}$

The high-T_c superconductor $YBa_2Cu_3O_{7-x}$ (YBCO) has a superconducting transition temperature T_c around 92 K, which is just above the boiling point of liquid nitrogen (LN_2). Therefore, YBCO can be used by cooling with LN_2, which is much cheaper than cooling with helium. The compound YBCO was synthesized in 1987 [8], soon after the Nobel Prize-winning discovery of high-T_c superconductivity in the cuprates by Bednorz and Müller [9]. Up to now, there is no consensus on the description of the high-T_c physical mechanism, unlike in the case of the so-called classical or low-T_c superconductors as found in the theory of Bardeen, Cooper and Schriever (BCS) [10].

YBCO is routinely made by PLD on several substrates. The most widely used substrate material is $SrTiO_3$(001), which is considered to adopt the perfect cubic perovskite ABO_3 structure, where A and B are cations of which the combined valence adds up to 6+. The bulk structure of YBCO is an oxygen deficient triple-perovskite, consisting of three tetragonal ABO_{3-x} blocks, stacked in their c-axis direction. On the STO(001) surface, YBCO can be grown with a c-axis orientation, which results in an in-plane lattice mismatch of ~0.2 % between the substrate and film. In turn, the lattice mismatch introduces a tensile strain in the film. Strain is omnipresent in thin film hetero-epitaxy and is known to influence the growth, by effecting the surface diffusion, and the film's physical properties. In bulk compounds, the superconducting transition temperature depends on the pressure, i.e. at very high pressures many compounds show enhanced values of T_c. For thin films this means that one may expect enhanced T_c's for compressive strain, which has been seen in the case of $La_{1.9}Sr_{0.1}CuO_4$ films on $SrLaO_4$ substrates [11]. Therefore also tensile strain in thin films is expected to lower T_c. Indeed, for YBCO this is observed for very thin films, in which strain is not yet relaxed [12]. With the YBCO film becoming thicker, strain is relaxed, and for films thicker than about 10 nm, values for T_c close to those of the bulk are found. An intriguing question that arose when attempts were made at growing one unit cell thin YBCO films, is that such samples showed no superconductivity anymore. It turned out, in fact, that the initial growth of YBCO does not proceed in a unit cell fashion, but rather in perovskite-block fractions of the bulk unit cell [13, 14]. Furthermore, cationic disorder in the interface region also results in a structure different from the bulk. Exchange of Y and Ba has been observed [14], thereby leaving the perovskite $ACuO_3$ building blocks intact, but it has also been found that A/B exchange could take place in (surfaces of) thin copper-oxide films [15]. The former type of disorder is expected to have a minor influence on the superconducting properties, which are believed to take place in the CuO_2 sheets. The latter type of cationic disorder would disturb severely the planar Cu–O structure, and is therefore expected to have a much more detrimental impact.

2.3.1.3 Experimental Set-Up

For the purpose of in-situ surface X-ray diffraction during PLD, a sample environment, which allows for controlled atmosphere is necessary. Typical deposition parameters include substrate temperatures up to 1200 K and

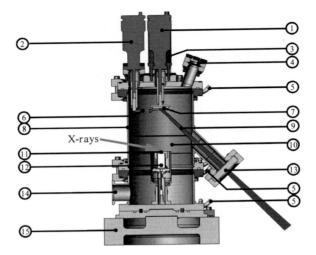

Fig. 2.2 Cross-sectional view of the portable PLD chamber. The height from base plate to top lid is 260 mm, the diameter of the middle part is 140 mm. *1* Feedthrough for target spindle. *2* Feedthrough for pre-ablation shutter. *3* Electro-motor for target spindle. *4* Viewport. *5* Water-cooling connections. *6* Pre-ablation shutter. *7* Target. *8* Aluminium chamber wall. *9* Laser beam. *10* Thinned part of aluminium wall (0.5 mm). *11* Substrate position. *12* Resistive wire heater embedded in nickel heater block. *13* Fused silica laser window. *14* Vacuum pump connection. *15* Base plate (d = 150 mm)

oxygen atmospheres ranging from 10^{-6} mbar during deposition up to 1 bar during post-growth annealing. At the same time, there need to be windows that are X-ray and laser transparent. Figure 2.2 shows schematically a portable PLD chamber that has been constructed for experiments at the ESRF.

The aluminum X-ray window allows 270° access to the sample. Part of the X-ray window is blocked by the laser access, which consists of a quartz window through which the focused laser beam impinges on the target. In order to prevent deposition during cleaning of the target surface with the first couple of laser shots, a shutter can be rotated in between the substrate and target. The target is mounted on a motorized rotational feed through, which allows for evenly distributing the ablation spots caused by the focused laser over a ring. This prevents from hitting the same part with each laser pulse.

The chamber is mounted on the diffractometer, whereby the laser is aligned in such a way that the intensity of a specularly reflected beam can be monitored during deposition. Since the laser window is fixed to the chamber and with that to the substrate, movement of the sample diffractometer angles rotates the laser window out of the laser beam path. In an upgraded version of the chamber, the substrate is mounted on a feedthrough, which allows for sample movement without changing the position of the laser window [16]. This improvement facilitates the alignment of the sample to monitor any point in reciprocal space during deposition without the need to change the laser alignment.

2.3.1.4 In-situ Surface X-ray Diffraction During Thin Film Growth

Within the framework of kinematical X-ray scattering, surface diffraction is described by Crystal Truncation Rods (CTRs) [17, 18]. The presence of a surface introduces rods of intensity along the surface normal in the diffraction pattern. The intensity at different reciprocal space points hkl whereby l, which is no longer an integer, runs along the CTR is calculated by summing up scattering contributions of unit cell layers, which end at the surface, giving:

$$F_{CTR}^{hkl} = \sum_{j=-\infty}^{0} F_{bulk}^{hkl} e^{i2\pi lj} e^{j\alpha} = \frac{F_{bulk}^{hkl}}{1 - e^{-i2\pi l} e^{-\alpha}} \tag{2.1}$$

where F_{bulk} is the structure factor of the bulk unit cell and α is the attenuation per unit cell layer. As usual, the intensity is related to the structure factor by $I \propto |F|^2$. The CTR is confined in reciprocal space to lines of intensity which are δ-like functions along h and k, and continuous along l. In the case of a film growing on a foreign substrate, the time-dependent scattering can be written as

$$F_{tot}^{hkl}(t) = F_{CTR}^{hkl} + \sum_{j=0}^{N} \theta_j(t) F_{film}^{hkl} e^{i2\pi lj} \tag{2.2}$$

where F_{film} is the structure factor of the film's unit cell and θ_j is the time-dependent coverage of the film's j-th unit cell layer.

Equation (2.2) takes the form of an oscillating intensity in the case of layer-by-layer growth and has been extensively examined for all kinds of different growth evolutions [19–21]. The detailed shape, in particular the periodicity of the oscillations, depends on the point hkl in reciprocal space and on the atomic structures involved. A simple form for the oscillations is seen at the anti-Bragg points such as ($00\frac{1}{2}$), where the intensity oscillates between two values with a period corresponding to the growth of one unit cell layer. In fact, (2.2) describes the elastic scattering near the CTR, which is centred at point (h, k). As two-dimensional islands appear on the surface, the associated diffuse scattering emerges, whereby these signals peak (Δh, Δk) from the CTR. The diffuse scattering can be calculated by Fourier transform of the correlation function, a two-point probability function, which gives the probability of finding two islands a certain distance apart [22]. The correlation function may include the island size-distribution and shape. In the case of isotropically shaped islands evenly distributed over the surface, the diffuse scattering manifests itself as a hollow cylinder around the CTR [22, 23]. The radius of the cylinder is related to the average distance between islands, whereas the width of the cylinder's jacket is inversely proportional to the average island size. The diffuse scattering can then be observed as additional peaks around the central CTR signal.

In the case of step-flow growth the surface diffusion is large enough that all the atoms reach step edges before nucleating. The surface morphology does not change upon deposition of fresh material and the long-range order as present on

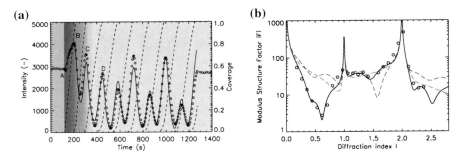

Fig. 2.3 a Intensity oscillations measured at the YBCO anti-Bragg point during PLD. As the film grows thicker in time, smooth and rough morphologies are encountered, which are marked by intensity maxima and minima respectively. Although the growth rate is constant, the first two smooth layers (**b** and **c**) are deposited in 2/3 the time needed for subsequent layers. **b** Specular CTR measured after completion of the first smooth layer. Shown are the data (*open circles*), the best fit (*black solid line*) and two model calculations of a single perovskite block (*red dashed*) and a double perovskite block (*blue dashed*). The best fit is the average, taking into account the phases, of the double and single perovskite block structure

the substrate surface is maintained. Usually, the steps are separated several tens of nanometers (several hundred atomic distances) and are therefore uncorrelated. The resulting scattering in this case can be described by summing the scattered intensities of the already overgrown part of the crystal and the uncovered part by:

$$I^{hkl} \propto \left(1 - \theta_j(t)\right)\left|F_{CTR}^{hkl} + F_{film}^{hkl}\right|^2 + \theta_j(t)\left|F_{CTR}^{hkl} + F_{film}^{hkl} + F_j^{hkl}\right|^2 \qquad (2.3)$$

where F_{CTR} is again the substrate contribution, F_{film} the scattering of the hitherto grown film, and F_j the structure factor of the j-th (presently growing) layer having θ coverage. Equation (2.3) gives a constant intensity in the case of homoepitaxy, which is understood because neither the crystal structure nor the roughness change during step-flow.

2.3.2 Experimental Results

YBCO thin film growth studies have been performed at the ESRF. The main goal was to follow the structure evolution during the heteroepitaxial growth on STO, a lattice-matched substrate (misfit ~0.2 %). By monitoring the YBCO anti-Bragg point, the expected intensity oscillations should have a period which corresponds to the YBCO unit cell. Figure 2.3 shows the resulting intensity oscillations, which clearly show a transition in period after the second intensity maximum. Qualitatively, this can immediately be interpreted as arising from a change in growth unit, which is smaller than the complete bulk YBCO unit cell. In another experiment, an ultra-thin layer was deposited, after which a complete data set

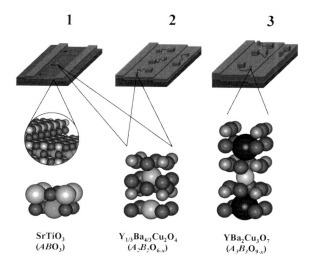

Fig. 2.4 Three-step structural evolution during the initial growth of YBCO on STO. In step 1, double perovskite $(Y,Ba)CuO_2$ blocks grow in a step-flow fashion. Then, in step 2, the same double perovskite blocks grow layer-by-layer thereby forming 2D islands on the surface. Finally, in step 3, complete cationically stoichiometric YBCO layers start to grow in a layer-by-layer fashion. This growth unit and mode persist up to film thicknesses of several tens of nanometers

consisting over several crystal truncation rods was measured. Unfortunately, this ultra-thin layer proved to be metastable, and only the specular CTR measured directly after deposition was usable.

The picture that emerges from these in-situ measurements is summarized in Fig. 2.4. During the initial growth, the crystal structure goes through an intermediate, metastable state before bulk YBCO starts to form. Although the dynamics are characterized by layer-by-layer growth, the very first layer proceeds in a step-flow fashion. This can be understood as YBCO sticking better onto itself than onto STO. The factors that are of influence on the surface diffusion in the heteroepitaxial case include bonding of the foreign species on the substrate and the additional energy barrier for diffusion over step-edges, the so-called Ehrlich-Schwöbel barrier [24, 25]. Both these energy terms change during the course of deposition of the very first layer. Furthermore, in the case of YBCO the change of growth unit further adds to the energy balance. The experiments described here have been repeated for different STO substrates, all with different miscut angles. Those substrates which contained terraces of the order of 200 nm and larger, showed that the very first layer grows in a mixed layer-by-layer and step-flow mode. Substrates with terrace sizes of about 100 nm and smaller showed only step-flow growth for the very first layer. YBCO growth on all substrates eventually results in layer-by-layer growth. This is an indication that indeed the surface diffusion length for the very first layer growing on the foreign STO substrate is significantly larger than for subsequent layers.

2.3.3 Deeply Buried Interfaces

Interfaces are omnipresent in real, finite systems and are very often of crucial importance to the physical properties, in particular when the volume to surface ratio decreases as in nanosystems. Whereas vacuum/crystal interfaces have been studied very extensively over the past decades, buried interfaces have not because of the limited experimental techniques that are suited for their study. One possibility is the use of high-energy X-rays that can reach and leave the interface of interest without being attenuated too much [26]. The most straightforward technique that can then be used to study the structure of interfaces is X-ray Reflectivity. In the following section this X-ray technique will be briefly introduced, after which a dedicated set-up for high-energy X-ray reflectivity studies is presented.

By the use of this set-up, available at the high-energy beamline of the European Synchrotron Radiation Facility and beamline P07 at Petra3@DESY, several studies have been done so far. Molten metals, like indium and lead, in contact with Silicon, show five-fold in-plane ordering and an unusual densification of the liquid at the interface [27, 28]. The in-plane ordering is the first experimental observation of five-fold symmetry, which is expected to appear in liquids. The densification is interpreted as charge being transferred across this Schottky junction, thereby shrinking the ionic radii of the metals. Another study unveiled the hydrophobic gap between water and octadecyl-trichlorosilane, which forms a self-assembled monolayer on Silicon wafers capped with its native oxide [29]. Obtaining a molecular-scale picture of the interfaces between hydrophobic materials and water is of importance to understand their dissolution properties. Recent instrumentation developments will allow for studies of interfaces at elevated pressures [30]. A special high-pressure cell is developed, which enables to study for example the water-methane interface, whereby the gas is pressurized up to 100 bar. These developments open up hitherto completely unexplored parts of the phase diagrams of various kinds of systems and will also allow studying the kinetics of a wide variety of processes.

2.3.3.1 X-ray Reflectivity

X-rays being electro-magnetic radiation, their properties are readily described by the Maxwell equations, which are at the basis of ordinary optics. However, due to the fact that the frequency of X-rays is very high there are some differences with visible light worth of mentioning.

The refractive index in the case of X-rays can be written as,

$$n = 1 - \delta + i\beta \tag{2.4}$$

whereby the parameters δ and β describe the dispersion and absorption, respectively, by

$$\delta = \frac{\rho \, r_0 \lambda^2}{2\pi} \tag{2.5}$$

Fig. 2.5 Directions of reflected and refracted X-rays at an interface whereby $n_0 < n_1$

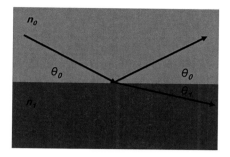

and

$$\beta = \frac{\lambda \mu}{4\pi} \tag{2.6}$$

with ρ the electron density of the medium, r_0 the classical electron radius, λ the wavelength and μ the linear absorption co-efficient. With the real part of the refractive index being slightly smaller than unity (δ is typically of the order 10^{-6}), X-rays 'bend' in the opposite direction compared to visible light when traveling into a denser medium. Figure 2.5 illustrates this by showing the directions of the reflected and refracted X-rays at an interface, whereby the beam travels from a less (n_0) to a more (n_1) dense medium. This results in the angle θ_1 being smaller than θ_0, which can also be calculated using Snell's law:

$$n_0 \cos (\theta_0) = n_1 \cos (\theta_1) \tag{2.7}$$

Total external reflection occurs for incoming angles θ_0 smaller than the critical angle $\theta_c \approx \sqrt{2\delta}$. For such small angles, the reflectivity R, which is defined as the ratio between the incoming (I_0) and reflected (I) intensity, becomes unity. The reflectivity curve as a function of scattering angle is described by the Fresnel equation, of which the shape shows a plateau up to θ_c and is inversely proportional to θ^4 for larger angles.

X-ray reflectivity (XRR) is useful to study the difference in electron density across an interface, or multiple interfaces which appear when thin films are present. The laterally averaged electron density $\rho(z)$, where z is defined along the interface normal, can be extracted with near-Ångström resolution from X-ray reflectivity profiles using several well established schemes [31].

Extending XRR measurements to the high X-ray energy regime puts stringent demands on the sample positioning and stability. The value of θ_c and the range over which the XRR curve is recorded scale approximately linearly with the wavelength, which means that angular resolution by which the sample has to be moved needs to match those. Figure 2.6 shows an XRR curve taken off the Si/SiO$_2$/vacuum interface using an X-ray energy of 72 keV. The critical angle is only $0.032°$ and already around a scattering angle of $0.5°$ has the intensity decreased by approximately eight orders of magnitude. The sample that was used for these

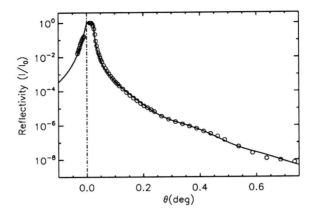

Fig. 2.6 X-ray reflectivity curve of the Si(100)/SiO$_2$(19Å)/vacuum interface. The *black line* is a fit to the data, which were taken at an X-ray energy of 72 keV. The critical angle is 0.0327°, but is well resolved, just as the faint oscillation due to the density difference between the native oxide and the silicon substrate. The high energy allows to measure through the Silicon, which was done here for negative values of θ

measurements was a polished Silicon disc of 20 mm diameter. At an X-ray energy of 72 keV, the transmission through the Si is about 30 %. As a result, it is also possible to measure the reflectivity curve on the 'other side', which means that the incoming and scattered beams travel through the Si. With the choice of angles here, this situation arises for negative values of θ. The few points measured in this geometry clearly show the absence of a plateau of total external reflection when approaching very small scattering angles, and the asymmetric shape of the curve around $\theta = 0$ becomes visible. Would the Si substrate be brought into contact with a medium of higher density than the plateau of total reflection would appear and the transmission through the Si would be still high enough to measure an appreciable part of the curve.

2.3.3.2 High-Energy Microdiffraction

A special set-up for high-energy X-ray Reflectivity and other diffraction studies requiring a micrometre-sized beam has become available at the European Synchrotron Radiation Facility [32]. A special diffractometer (see Fig. 2.7) allows for the required sample positioning and stability. The use of focusing by compound refractive lenses, rendering high-energy X-ray micro beams results in a much improved signal to noise ratio. Hereby it is now possible to study deeply buried interfaces by penetrating several centimetres of Si. The set-up can be used for X-ray reflectivity measurements and in-plane diffraction to follow the evolution of the Bragg peaks of the growing film or nanocrystals. Furthermore, experiments can be done using a so-called liquid monochromator, which tilts the incoming beam with respect to the sample surface.

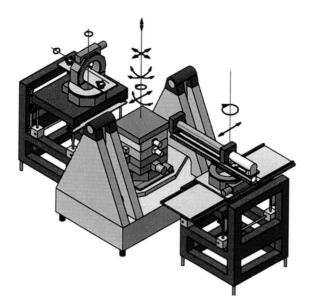

Fig. 2.7 Sketch of the HEMD set-up at the ESRF. The liquid monochromator composes a four-circle diffractometer, on which an arm with two silicon crystals are mounted (*left*). The primary beam is deviated by a Bragg reflection of the first crystal, after which it hits the second one. The second crystal is also in diffracting position. The original primary beam and the two reflected beams make up an isosceles triangle. By rotating the crystals around the original primary beam direction, the beam reflecting off the second crystal is tilted vertically without changing the spot where it intersects the horizontal plane. The sample is mounted on a tower of several high precision positioning motors, which in turn are all in mounted on a granite swing (*middle*). The detector sits at the end of a collimator, which can be rotated and translated in order to define the horizontal angle. By moving separately the front and rear legs of the table, the vertical detector angle is adjusted (*right*)

2.4 X-ray Free Electron Lasers

As explained in more detail in Chap. 7 on detectors, X-ray Free-Electron Lasers (XFEL) are the fourth and next generation of X-ray sources, showing an increase in peak brilliance of nine orders of magnitude as compared to the third generation storage rings. It is obvious that this will make new experiments as well as new science possible. Although the first X-ray Free-Electron Laser sources are now coming online and are producing first science, it is not yet clear which scientific fields will profit most from these fundamentally new sources, but it seems reasonable to expect that in-situ scattering experiments will be part of the scientific portfolio at these sources. The peak brilliance is the most impressive parameter of Free-Electron Lasers due to the short bunch length, but also the average brilliance, a more common parameter used for storage ring sources, will see an increase of four orders of magnitude as compared to existing third generation storage rings.

The great promise and excitement produced by these new sources is maybe best indicated by the number of FELs under construction throughout the world. The first short-wavelength FEL operational was FLASH at DESY in Hamburg, operating in the XUV range, and producing exciting new science since 2006. The LCLS at SLAC in the USA, is first hard X-ray Free-Electron Laser and produced first laser light in 2009. A number of new sources will come online and enter user-service mode in the years to come, notably The FERMI project in Italy, and the SPring-8 Angstrom Compact Free Electron Laser in Japan. In 2015 the European XFEL in Hamburg and in 2017 the SwissFEL at PSI will become operational.

2.4.1 Source Parameters

X-ray Free-Electron Lasers are fundamentally different sources than synchrotron storage rings. The principle idea behind the FELs is to compress the electron bunches as much as possible, and to use extremely long undulators, of 100 m or more. Due to the high electron density inside the bunch, the electrons will experience their own radiation field while transferring the undulator, and as a result so-called micro-bunches will start to form. The electrons inside a micro-bunch will radiate all in phase and by that give an exponential increase in emitted intensity. This so-called Self-Amplified Spontaneous Emission, or SASE, only happens with highly compressed electron bunches and extremely long undulators, which is not possible with equilibrium sources like Synchrotron storage rings. The expected parameters for the European XFEL are given in Table 2.1, as defined in the Technical Design Report in 2006 [33]. It should be pointed out that these parameters are constantly evolving with increasing know-how in accelerator science.

The most relevant parameters are probably the pulse length of less than 100 fs, as compared to 100 ps for storage rings, and the 10^{12} photons per pulse. Furthermore, the X-ray beam is fully laterally coherent. The tremendous peak power of 20 GW inside a 100 fs pulse will open up many atomic physics and plasma physics experiments, where the focused beam will turn any sample into a plasma or warm dense matter. However, this does not mean that in-situ scattering experiments are excluded as explained in the following paragraph.

2.4.2 Potential for In-situ Experiments at FELs

One of the pre-requisites to perform in-situ scattering experiments on dynamic processes is that the probe, in this case the X-ray beam, should not significantly modify the system under study. Since the cross-section for X-ray scattering is relatively low, this is almost always the case in storage ring based experiments. In contrast, the FEL beam will be powerful enough to significantly alter, or even destroy, the sample when focused to a small spot. Therefore, the in-situ scattering experiments will use either unfocussed and apertured beams or reduce the flux by using a high-resolution monochromator and a small bandwidth, or both. Another parameter that has to be taken into account when designing the in-situ experiments

Table 2.1 Source parameters for the European XFEL as defined in the 2006 Technical Design Report [33]

	Unit	SASE I		SASE 2		SASE3	
Electron energy	GeV	17.5	17.5	17.5	17.5	17.5	10
Wavelength	nm	0.1	0.1	0.4	0.4	1.6	4.9
Photon energy	keV	12.4	12.4	3.1	3.1	0.8	0.25
Peak power	GW	20	20	80	80	130	150
Average power	W	65	65	260	260	420	490
Photon beam size (FWHM)	μm	70	85	55	60	70	90
Photon beam divergence (FWHM)	μrad	1	0.84	3.4	3.4	11.4	18
Coherence time	fs	0.2	0.22	0.38	0.34	0.88	1.4
Spectral bandwidth	%	0.08	0.08	0.18	0.2	0.3	0.65
Pulse duration	fs	100	100	100	100	100	100
Photons per pulse	#	10^{12}	10^{12}	1.6×10^{13}	1.6×10^{13}	1.0×10^{14}	3.7×10^{14}
Average flux	#/s	3.0×10^{16}	3.0×10^{16}	4.8×10^{17}	4.8×10^{17}	3.1×10^{18}	1.1×10^{19}
Peak brilliance	B	5.0×10^{33}	5.0×10^{33}	2.2×10^{33}		5.0×10^{32}	1.0×10^{32}
Average brilliance	B	1.6×10^{25}	1.6×10^{25}	6.5×10^{24}		1.4×10^{24}	2.8×10^{23}

at the European XFEL is the special time structure, with pulse trains of 0.6 ms long containing 2700 pulses (220 ns pulse spacing), followed by a 99.4 ms long inter-train interval (see Fig. 7.19, Chap. 7). This means that various timescales will be accessible: 100 fs using single pulses, 200 ns using consecutive pulses, 600 µs using the entire bunch trains and 100 ms or longer using consecutive pulse trains. This is a significant difference with other FELs that use non-superconducting accelerator technologies and produce equally spaced single pulses with 60 to 120 Hz repetition rates. For these sources only the femto-second and second timescales will be accessible.

Besides the classical in-situ scattering experiments already developed at third generation storage rings, Coherent Diffraction Imaging (CDI) and X-ray Photon Correlation Spectroscopy (XPCS), both using the coherence of the beam, seem to be particularly promising.

2.4.2.1 Coherent Diffraction Imaging (CDI)

Coherent Diffraction Imaging (CDI) has rapidly developed as a form of lensless X-ray microscopy, and was pushed by the fact that oversampled diffraction patterns can be inverted to obtain real space images, as pointed out by Sayre already in 1952 [34]. When recording the diffracted intensities, the phase information is lost, but, due to the Fourier relationship, still embedded in the intensity profile. This phase information can be recovered, provided the image is sufficiently oversampled. The inversion from reciprocal space to real space has been proven to be unique (accept for some pathological cases), and various computational methods and programs have been developed, using iterative procedures and general constraints like the finite support of the sample. Due to these tools and the availability of coherent X-ray beams, Coherent Diffractive Imaging has provided insight in the structure of materials on the nanometer scale. It is to be expected that the high degree of coherence of the X-ray FELs will give a significant push in this field. First experiments at the LCLS have clearly demonstrated the promises of coherent diffraction imaging [35, 36]. Pioneering work by Robinson [18] showed that the CDI technique can yield nanometer scale information on strains of small particles in contact with a surface or other particles [37]. Using the time structure of the European XFEL on can follow such interfacial strain states as function of time, or process parameters.

2.4.2.2 X-ray Photon Correlation Spectroscopy (XPCS)

Another technique that will profit from the increased coherence of the beam, is X-ray Photon Correlation Spectroscopy (XPCS). XPCS is the extension of the well established visible light photon correlation spectroscopy into the X-ray regime. The use of X-rays in stead of visible light allows the study of otherwise opaque systems, and gives access to nano-meter length scale dynamics. The lack of coherent flux at third generation storage rings has been the major obstacle for wide application of the technique, therefore a major expansion of XPCS can be expected at the XFELs. In XPCS the sample is illuminated by the fully coherent beam and a speckle pattern is created, either in the forward small-angle scattering region, or in the wide-angle diffraction region. The intensity fluctuations in time of individual speckles are then recorded. Since this can be done at different momentum transfers, information of the dynamics at different lengths scales is obtained.

One of the boundary conditions for XPCS is that the probing X-ray beam should not significantly alter the state of the sample under study. This means that the incoming beam cannot be focused to a small spot, since this would cause a considerable temperature increase even in dilute samples. Consequently, in order to profit from the increased coherent flux, one is restricted to using relatively large samples. More examples of the possibilities for XPCS experiments at the European XFEL can be found in the Technical Design Report [33].

2.5 Summary

In this chapter we have shown that X-rays have certain advantages over charged particles as probe for in-situ studies. These advantages stem mainly from the low scattering cross section for X-rays, which makes the theoretical analysis of the scattering intensities much easier, and allows using sophisticated sample chambers, as well real-life experimental conditions, like atmospheric pressure. Both the experimental and the theoretical concepts are therefore well developed and mature. A disadvantage of the low scattering cross-section is the corresponding inefficient use of the available flux, which needs to be compensated by powerful sources, like synchrotrons. This in turn makes these techniques less accessible, especially for students, than laboratory-based techniques.

The power and possibilities of in-situ X-ray diffraction are illustrated with two examples. The first example is the study of the initial growth of thin films inside Pulsed-Laser Deposition chambers. The second example is the study of deeply buried interfaces. Both studies are only possible with medium to high energy X-rays.

The final part of this chapter presents the upcoming X-ray Free-Electron Laser sources (XFEL), which give and increase in peak brilliance of nine orders of magnitude over existing third generation synchrotron sources. In addition, the beam will be fully laterally coherent, allowing for X-ray Photon Correlation Spectroscopy (XPCS) and Coherent Diffractive Imaging (CDI) experiments. The full potential of these sources for in-situ experiments will only become clear over the next years, with the first XFELs going into operation.

References

1. H.F. Poulsen, *Three-Dimensional X-Ray Diffraction Microscopy: Mapping Polycrystals and their Dynamics*, ISBN 3-540-22330-4, (2004)
2. S.B. Ogale (ed.), *Thin Films and Heterostructures for Oxide Electronics* (Springer + Business Media, New York, 2005)

3. H. Karl, B. Stritzker, Phys. Rev. Lett. **69**, 2939 (1992)
4. A.J.H.M. Rijnders, G. Koster, D.H.A. Blank, H. Rogalla, Appl. Phys. Lett. **70**, 1888 (1997)
5. G. Eres, J.Z. Tischler, M. Yoon, B.C. Larson, C.M. Rouleau, D.H. Lowndes, P. Zschack, Appl. Phys. Lett. **80**, 3379 (2002)
6. A. Fleet, D. Dale, Y. Suzuki, J.D. Brock, Phys. Rev. Lett. **94**, 036102 (2005)
7. H.-H. Wang, A. Fleet, J.D. Brock, D. Dale, Y. Suzuki, J. Appl. Phys. **96**, 5324 (2004)
8. M.K. Wu, J.R. Ashburn, C.J. Torng, P.H. Hor, R.L. Meng, L. Gao, Z.J. Huang, C.W. Chu, Phys. Rev. Lett. **58**, 908 (1987)
9. J.G. Bednorz, K.A. Müller, Z. Phys. **B64**, 189 (1986)
10. J. Bardeen, L.N. Cooper, J.R. Schrieffe, Phys. Rev. **108**, 1175–1204 (1957)
11. J.-P. Locquet, J. Perret, J. Fompeyrine, E. Mächler, J.W. Seo, G. van Tendeloo, Nature **394**, 453 (1998)
12. H.Y. Zhai, W.K. Chu, Appl. Phys. Lett. **76**, 3469 (2000)
13. T. Haage, J. Zegenhagen, H.-U. Habermaier, M. Cardona, Phys. Rev. Lett. **80**, 4225 (1998)
14. V. Vonk, K.J.I. Driessen, M. Huijben, G. Rijnders, D.H.A. Blank, H. Rogalla, S. Harkema, H. Graafsma, Phys. Rev. Lett. **99**, 196106 (2007)
15. X. Torrelles, C. Aruta, A. Fragneto, I. Maggio-Aprile, L. Ortega, F. Ricci, J. Rius M. Salluzzo, U. Scotti di, Phys. Rev. B **70**, 104519 (2004)
16. P. Tinnemans, DyScO3 thin film substrates—an X-ray diffraction study, PhD Thesis, Radboud University Nijmegen, The Netherlands (2012)
17. R. Feidenhans'l, Surf. Sci. Rep. **10**, 105 (1991)
18. I.K. Robinson, *Handbook on Synchrotron Radiation*, vol. 3, chap. 7. (North-Holland, Amsterdam, 1991), pp. 221–266
19. C.S. Lent, P.I. Cohen, Surf. Sci. **139**, 121–154 (1984)
20. P.I. Cohen, G.S. Petrich, P.R. Pukite, G.J. Whaley, A.S. Arrott, Surf. Sci. **216**, 222–248 (1989)
21. P.R. Pukite, C.S. Lent, P.I. Cohen, Surf. Sci. **161**, 39–68 (1985)
22. E. Vlieg, J.F. van der Veen, S.J. Gurman, C. Norris, J.E. MacDonald, Surf. Sci. **210**, 301–321 (1989)
23. H.A. Van der Vegt, W.J. Huisman, P.B. Howes, E. Vlieg, Surf. Sci. **330**, 101–112 (1995)
24. G. Ehrlich, F. Hudda, J. Chem. Phys. **44**, 1030 (1966)
25. R.L. Schwöbel, E.J. Shipsey, J. Appl. Phys. **37**, 3682 (1966)
26. H. Reichert, V. Honkimaki, A. Snigirev, S. Engemann, H. Dosch, Phys. B: Condens. Matter. **336**, 46–55 (2003)
27. H. Reichert, O. Klein, H. Dosch, M. Denk, V. Honkimäki, T. Lipmpmann, G. Reiter, Nature **408**, 839 (2000)
28. H. Reichert, M. Denk, J. Okasinski, V. Honkimäki, H. Dosch, Phys. Rev. Lett. **98**, 116101 (2007)
29. M. Mezger, H. Reichert, S. Schöder, J. Okasinski, H. Schröder, H. Dosch, D. Palms, J. Ralston, V. Honkimäki, PNAS **49**, 18401 (2006)
30. F. Venturini, S. Schöder, W.F. Kuhs, V. Honkimäki, H. Reichert, H. Schober, F. Thomas, J. Synchrotron Rad. **18**, 251 (2011)
31. M. Tolan, *X-ray Scattering from Soft Matter Thin Films* (Springer, Berlin, 1999)
32. V. Honkimäki, H. Reichert, J.S. Okasinski, H. Dosch, J. Synchrotron Rad. **13**, 426 (2006)
33. M. Altarelli et al., *Technical Design Report of the European X-ray Free Electron Laser*, ISBN 978-3-935702-17-1 (2006)
34. D. Sayre, Acta Cryst. **5**, 843 (1952)
35. H. Chapman et al., Nature **470**, 73–77 (2011)
36. M.M. Seibert et al., Nature **470**, 78–81 (2011)
37. M.A. Pfeifer, G.J. Williams, I.A. Vartanyants, R. Harder, I.K Robinson *Nature* **442**, 63–66 (2006)
38. Cross-sections were taken from the following databases: www.ioffe.rssi.ru, www.nndc.bnl.gov, www.esrf.eu/computing/scientific/xop2.1

Chapter 3
In-situ Transmission Electron Microscopy

Xiao Feng Zhang

Abstract Transmission electron microscopy (TEM) has hit a significant milestone of sub-angstrom resolution. On one hand, electron microscopists and materials scientists are enjoying the highest TEM spatial resolution ever attainable; on the other hand, study of materials in a steady state is hard to meet the increasing demand in new application fields such as nanocatalysts, nanocrystal growth, nanoelectronics, nanosensors, and nanomechanics in which size effect and structural or property responses to stimuli from the surrounding environment are key information to learn. Special attention is thus paid to in-situ TEM. A great deal of effort in developing and improving electron microscopes and specimen holders have resulted in unprecedented progresses in attaining insight into materials in dynamic environments. In many ways, transmission electron microscopes are now functionalized as workstations or nanoscale labs rather than just imaging tools. In this chapter, various types of in-situ TEM technologies are introduced accompanied by application examples. In parallel to the sub-angstrom breakthrough made by the aberration-corrected TEM, atomic resolution is now emphasized in advanced in-situ TEM, advancement on this aspect will be discussed together with other important notes and further challenges.

3.1 What is the In-situ TEM and Why It Is Important

One of the milestones in materials characterization technology in the 20th century was the invention of the transmission electron microscope in 1931. Among many advantages of transmission electron microscopy (TEM) compared with other imaging technologies,

X. F. Zhang (✉)
Nanotechnology Systems Division, Hitachi High Technologies America, Inc.,
5960 Inglewood Drive, Suite 200, Pleasanton, CA 94588, USA
e-mail: xiao.zhang@hitachi-hta.com

X. F. Zhang
Hitachi High Technologies Corporation, Tokyo 105-8717, Japan

A. Ziegler et al. (eds.), *In-situ Materials Characterization*,
Springer Series in Materials Science 193, DOI: 10.1007/978-3-642-45152-2_3,
© Springer-Verlag Berlin Heidelberg 2014

a distinctive and very unique one is the twofold structural information obtainable on the same instrument. Electron diffraction tells structural information in the reciprocal space just like what X-ray and neutron diffractions tell, and the real space TEM images reveal morphologies and atomic lattice structures of objects. Five decades later, instrumentation and TEM have been well advanced and also combined with spectrometers for chemical analysis. TEM has become indispensable in materials science especially in the study of individual sub-micron size objects (such as nanomaterials, precipitates, grain boundaries, biological cells, and molecules) and locally deformed structures (dislocations, stacking faults, and strained lattice). Because of the technical feasibility, conventional TEM has been focusing on stable structures in the electron microscope vacuum and at ambient temperature. However, since the main themes of today's materials science are moving to nanomaterials and energy-related materials, researches in catalysis, nanocrystal growth, solid/liquid/gas interactions, and gas sensing are largely stimulated. Accordingly, the demand for in-situ TEM capabilities is growing fast. "The electron microscope should be a workstation rather than only an imaging tool", this comment from Dr. David C. Joy of the Oak Ridge National Laboratory/USA draws a picture for what is happening in the field of in-situ TEM.

What is in-situ TEM? While there might exist various definitions, a definition given in a report entitled "Dynamic in-situ electron microscopy as a tool to meet the challenges of the nanoworld" prepared for National Science Foundation of the USA may serve as an excellent explanation [1]. In this report, in-situ TEM was defined as *"Some form of stimulus is applied to a sample while it is observed in a TEM."* According to this definition, in-situ TEM has two important characters, *stimulus* and *real-time observation*. Stimulus, or say an external field, needs to be applied directly to a TEM specimen sitting in an electron microscope column. Typical external fields used for in-situ TEM include heating, cooling, electric, magnetic fields, as well as mechanical forces (tensile, compressive) and ion beam irradiation. To apply an external field to the specimen area and perform TEM observation simultaneously, one just needs, in most cases, a specially designed TEM specimen holder and a transmission electron microscope, which allows the in-situ TEM specimen holder to fit in. Also, a fast image recording system (for example a video recorder or TV rate CCD camera) is important. Sometimes, a modification of the transmission electron microscope is necessary. In this chapter, all these important facets of the in-situ TEM technology will be touched.

A question to ask is: why bother using in-situ TEM? Can't we simply treat materials ex-situ, say in a reaction chamber outside of an electron microscope and then move the treated sample back into the electron microscope for observation? One can certainly do the sample treatment/observation in this way, which is in fact one typical way of using TEM in the materials science field. But from ex-situ TEM experiments, only the beginning and ending status of materials can be learned. What has happened during the middle stages is missed. Misinterpretations of experimental data, wrong assumptions, or incorrect derivations were indeed not rare in reality when real-time observation capabilities were not available. A simple analog is crime scene investigation. For investigators who see the result of a criminal event, it is also not so difficult to learn about the scene before the crime. In order to figure out what happened during the criminal activity, investigators would have to collect all information they could dig out such as finger prints, blood samples, and residues. As electron microscopists and materials scientists, we are fortunate to have a state-of-the-art in-situ TEM to reveal structural evolution in materials in real time when it happens. Another angle to look at why we need the in-situ TEM is that the capability of applying external fields or forces to the TEM specimens actually turns the TEM specimen chamber into a miniaturized laboratory in which chemical reactions, structures, or physical properties can be activated or altered at micrometer to nanometer scales and all of the processes are monitored at high spatial and temporal resolutions.

3.2 A Brief History of In-situ Microscopy

The invention of the transmission electron microscope was driven, in part, by a couple of disease mysteries encountered in late 19th century. People attributed the causes to some invisible living objects much smaller than bacteria, which were viruses as we know today. The size of viruses is 0.1 mm or smaller which is beyond the resolving power of the optical microscope, therefore no one could prove the existence of viruses in the late 19th to early 20th century despite all of the disease analyses pointed to its existence. Microscopes with much higher resolution were urgently needed. Concomitant with the invention of the first transmission electron microscope in 1931, three high expectations were put on this new, yet unproven microscope: Higher resolution than that of the optical microscope, good imaging contrast on biological samples, and observation of live objects. With a continuous development in 80 years, TEM resolution has been increased by 1000 times, from about 50 nm in the beginning to 0.05 nm on today's most powerful microscopes. In the meantime, imaging contrast for biological samples has also been significantly enhanced by allowing adjustment of the electron accelerating voltage, the focal length of the objective lens, the size of the objective aperture, and recently introducing phase-plate technology.

In contrast to the achievements in high resolution and high contrast imaging, progress in seeking solutions for live observation of biological specimens remains unsatisfied. Although electron microscopes have obvious advantages over optical microscopes in terms of high resolution and chemical analysis, the optical microscope allows looking at specimens in their original states and environments such as in air or liquid solution. Interestingly enough, live microscopy, or in-situ microscopy, can be traced back to more than 300 years ago. In 1679, Antoni Van Leeuwenhoek published his observations on live microbes using his home-made, one-lens microscope, Fig. 3.1 [2]. The discovery of these "my little animals" (nickname given by Van Leeuwenhoek) virtually opened a door for human beings toward the microworld. Since then, observing and studying biological systems in their live state has long been the highest interest of biologists and the same enthusiasm was naturally carried over to TEM observations in the mid 20th century. However, a big hurdle was the vacuum condition in the electron microscope column. The high vacuum of $\sim 10^{-5}$ Pa in the specimen chamber prevents any live biological specimen to endure

Fig. 3.1 In 1679, Antoni Van
Leeuwenhoek published his
observations on live sperms
of a rabbit (*right*), using a
simple one-lens microscope
made by himself (*left*)

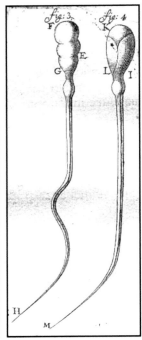

for a sufficient long time for electron microscopy study. The task was therefore
pretty clear: gas or liquid solution or both must be introduced into the specimen area
in the electron microscope. Such an ambient is not only beneficial to live observation of biological specimens, but also enables TEM studies on many other objects
such as catalytic process/mechanism, material growth, and electrochemical process
which require suitable gases or liquid solutions around specimens. In 1942, Ruska
was probably the first to report a low vacuum transmission electron microscope for
the purpose of providing a flexible environment to materials under TEM study [3].
In addition to the adjustable environment, external fields turned out to be crucial as
well to mimic the real world conditions inside electron microscope chambers. Today,
supplying gas or liquid solution to the specimen chamber of a transmission electron microscope is not a big challenge anymore. Also, by modifying TEM specimen
holders, various holder-based in-situ TEM technologies have already matured and
are widely used in materials science.

3.3 In-situ TEM Technologies

An important point the chapter author would like to make is that *to some extent,
in-situ TEM is all about specimen holders*. Depending on what external field is to be
applied to the specimen area, a corresponding specimen holder can be designed.

Most of the matured in-situ TEM technologies are based on elegant design of specimen holders like those used for in-situ heating, cooling, environmental, probe, straining, Lorentz, electron holography, and ion beam irradiation TEMs.

3.3.1 In-situ Heating and Cooling TEM

Perhaps the first in-situ heating TEM stage was reported in 1960 to observe annealing effects on dislocations in aluminum [4]. Nowadays, heating can be realized by placing a heating element at the tip of a TEM specimen holder. Depending on the heating element, heating mechanism, and sample fixing method, there are three typical types of commercially available in-situ heating TEM holders: furnace-heating holders, wire-heating holders, and membrane-heating holders. Joule heating of nanomaterials is also used in reported applications as will be described in Sect. 3.3.4.

At the tip of a furnace-heating holder, a heating filament embracing a 3 mm-diameter TEM specimen disk acts like an electric furnace. The thermal radiation heats the specimen, and is therefore an indirect heating mechanism. Cooling water is connected to the holder for use at above 500 °C. An embedded thermal couple measures temperature in the furnace cup. Because the 3 mm-diameter heating zone is "huge" and involves many components and supporting materials (e.g. TEM metal grid), the total thermal expansion effect causes a severe problem of sample drifting when changing temperature. It may take a few tens of minutes to seize the drifting, therefore taking high resolution TEM images in a short period of time requires a blessing. Obviously, this type of furnace-heating holder is good to perform low resolution in-situ TEM imaging [5–13]. As an application example, Fig. 3.2 shows a result of in-situ heating of a bilayer nanowire composed of a 20 nm-thick Cu layer and a 100 nm-thick SnO_2 layer. The bilayer nanowire was heated in a 300 kV transmission electron microscope. Because of the difference in thermal expansion coefficients between Cu and SnO_2, the bilayer nanowire curves toward the Cu side at room temperature as shown in Fig. 3.2a but becomes straight at 200 °C, Fig. 3.2b [9].

A demonstration of the in-situ wire-heating holder was reported by Kamino and Saka [14]. A spiral tungsten wire 20–50 mm in diameter is used to heat powder materials attached to the heating wire and is therefore based on a direct heating mechanism. The materials can be heated to as high as 1500 °C although the heating power is maintained at a low level, and the sample area being heated may be millions of times smaller than that of the furnace-heating holder. Cooling water is not required. The sample drifting rate due to the thermal expansion effect is tolerable ~10 min after temperature change, therefore atomic resolution at elevated temperatures is readily achievable as demonstrated in many published papers [14–23].

Figure 3.3 shows the kinematic growing process of a SiC crystal. Si particles on a graphite support were heated to 1500 °C in a 300 kV transmission electron microscope and the reaction between molten Si and graphite formed 6H-SiC.

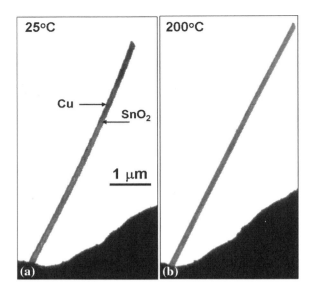

Fig. 3.2 TEM image of a 6.1 μm long bilayer nanowire composed of a 28 nm-thick Cu layer and a 90 nm-thick SnO$_2$ layer. The bilayer nanowire curves toward the Cu side at room temperature (**a**), and becomes straight at 200 °C (**b**)

Fig. 3.3 In-situ TEM observation of SiC growth at 1500 °C. An atomic monolayer is seen growing on the SiC surface. *Arrows* indicate the very forefront atoms at the moment

The three images present the successive moments in growth of an atomic layer on the SiC surface [15].

Using the same concept, heating holders with two wire heaters [16], three wire heaters [18], and multiple wire heaters with a gas spray nozzle [24] were also developed for doing in-situ evaporation deposition in electron microscopes. Figure 3.4 is an illustration of a double-heater holder. Precursor materials on the upper heater can be evaporated and deposited onto the substrate mounted on the lower heater. Solid-liquid-gas interactions can thus be studied.

The membrane-heating holder is rather new compared to the furnace-heating and wire-heating TEM holders. The key component of this type of holder is a heating

Fig. 3.4 Illustration of a double-heater in-situ TEM holder. The upper heater is for in-situ TEM evaporation deposition of materials onto a substrate positioned on the lower heater (Patent of Hitachi High Technologies Corporation, Japan)

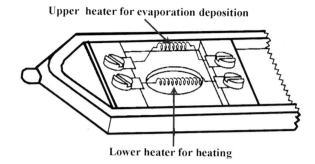

Upper heater for evaporation deposition

Lower heater for heating

device made from a conductive ceramic membrane suspended on a Si chip. Heating is very local and the heating power is small, resulting in a small sample drift rate at elevated temperatures. A very unique feature of the membrane-heating holder is the fast temperature change rate, as high as 10^6 °C/s is possible. This unique feature makes the holder a good choice for doing in-situ TEM thermal cycling experiments [25].

In addition to the three typical types of heating holders introduced above, a novel technology to be proven in applications is the in-situ environmental laser heating holder. The holder is assembled by a set of laser focusing and reflection optics. The laser beam heats a specimen without the presence of any heating elements. Temperature can be raised to beyond 2000 °C. The spot size of the laser beam and the irradiation position on the specimen can be varied. It is expected that the laser beam irradiation leads to a uniform concentric thermal expansion in the specimen, therefore minimizing uneven thermal stresses and ultimately a very low specimen drift rate is possible [26].

3.3.1.1 Chemical Analysis at Elevated Temperatures

Chemical analysis complements structural characterization is important for the in-situ heating TEM study because chemical processes are likely to occur when materials are heated. Energy-dispersive X-ray spectroscopy (EDS) and electron energy-loss spectroscopy (EELS) are two popular analytical methods associated with TEM. Because the EELS spectrometer is attached at the bottom of the electron microscope, its working condition is not influenced by field change in the microscope specimen chamber and is therefore ideal for in-situ heating TEM experiments. Kamino et al. demonstrated EELS analysis for SiO_2 at 700 °C [27]. In contrast, the EDS detector comes very close to TEM specimen (a few millimeters in distance), the infrared radiation from the heating zone may affect the detector, resulting in a dark current which in turn increases the noise in the X-ray signal detector. EDS analysis at elevated temperatures, e.g., 550 °C, was indeed reported [12]. At even higher temperatures, significant noise deteriorates the EDS energy resolution.

Figure 3.5 shows EDS spectra of a Si specimen in-situ heated in a transmission electron microscope equipped with a 30 mm^2 Si(Li) EDS detector. The full width

Fig. 3.5 EDS spectra
showing Si-K peaks at room
temperature and elevated
temperatures. Deteriorated
energy resolution is seen at
750 °C and above

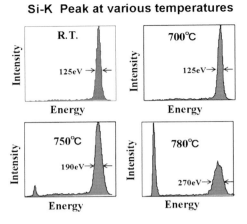

Si-K Peak at various temperatures

at half maximum (FWHM) point of the Si peaks is marked for the peak acquired
at room temperature, 700, 750, and 780 °C, respectively. It can be seen that the
FWHM value remains unchanged untill 700 °C, but broadens at 750 and 780 °C.
This test indicated a regular EDS performance until about 700 °C (for Si) but the
energy resolution deterioration was prominent at higher temperatures. Covering
the collimator of the EDS detector with a Beryllium window would solve this
problem at the expense of detection sensitivity for light elements. Another concern
is that the window in front of the EDS detector would become brittle if it is very
close to the heating zone, and the cycles between room and high temperatures may
eventually damage the window material and glue. The advice is, if the EDS detector is to be used in any in-situ heating TEM experiments, to consult with the EDS
system provider beforehand.

The opposite of the in-situ heating TEM is cryo TEM using liquid
nitrogen- (<100 K) or liquid helium- (4.2 K) cooled TEM holders. Although there
were experimental examples for in-situ cooling TEM to reveal materials' responses
to the low temperatures such as magnetic flux and vortices in superconductors at
below 10 K [28], cryo holders are actually widely used for stabilizing the structure, reducing sample contamination rate and beam damage rate, and maintaining
the temperatures required by frozen-hydrated biological samples (cells, tissues,
macromolecules, viruses). The latest applications are in cryo electron tomography
for three-dimensional imaging [29–31].

3.3.2 In-situ Gas Environmental TEM (ETEM)

From the engineering point of view, thermal and cryo fields are fairly easy to be
introduced into the specimen area in the transmission electron microscope because
the instrumental modification is limited to the TEM specimen holders. However, in
our real world, many physical and chemical processes take place in an environment

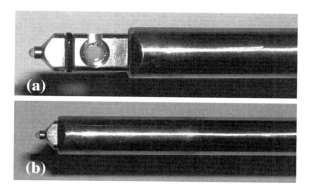

Fig. 3.6 An environmental transfer holder. **a** The specimen cup is in the Out position for specimen treatment in a reaction chamber outside of an electron microscope. **b** The specimen cup can be retracted and sealed in the rod of the holder to prevent air exposure during transferring from the external reaction chamber into the electron microscope column

filled with gas and/or liquid. For example, a catalytic process includes catalysts, gas, and a suitable temperature, which is, in most cases, higher than room temperature. It is clearly desirable to have an in-situ TEM technology that is capable of handling gas or liquid in the specimen chamber of the electron microscope.

As a simple and low cost solution, an environmental transfer TEM specimen holder may be a good compromise to partly meet the ETEM demand. Figure 3.6 shows a picture of such a holder. The tip of the holder where the TEM specimen locates is movable, so it can be retracted into the holder rod and sealed by an O-ring.

One can load a specimen into the holder, study the original state in an electron microscope, and then move the holder (with the specimen) to a reaction chamber outside the microscope. The specimen is treated in the reaction chamber under desired environmental conditions (gases, pressure, temperature, etc.). Before taking the holder out of the reaction chamber, the tip of the holder is retracted into the rod so that the specimen will not be exposed to air when transferred back into the electron microscope. The post-treatment structure can thus be studied. This type of transfer holder allows for a comparison between the material structures before and after material treatment but the middle links are unknown. An example is shown below.

Polycrystalline SiC was synthesized by hot pressing in the presence of aluminum, boron, and carbon additives [8]. About 1 nm-thick amorphous intergranular films were formed among SiC grains. After heat treatment at above 1000 °C for 3 days or more, the amorphous intergranular films seemed to disappear because the boundaries looked amorphous-free. In order to figure out where the amorphous intergranular films went, in-situ heating TEM was carried out. The grain boundary film with an amorphous structure shown in Fig. 3.7a was heated and monitored in a transmission electron microscope. The heating temperature was 1200 °C at which the sample was held for 25 h with the electron beam turned off to minimize the electron beam irradiation effects. Figure 3.7b shows the result. In fact, the grain boundary film remains at the same location but has been partially crystallized to form an Al-O-C structure.

Fig. 3.7 a High resolution
TEM image showing a
1 nm-thick amorphous
intergranular film between
two SiC grains. **b** After heat
treatment at 1200 °C for 25 h
in a transmission electron
microscope, crystallized
segments are observed along
the intergranular film

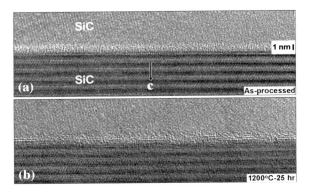

The in-situ TEM observation clarifies the newly formed Al-O-C crystalline structure contained in the "clean" boundary. The structure is hard to distinguish from the SiC matrix because it is similar to the SiC structure and has an epitaxial structural relationship with the SiC (0001) surface.

There is no doubt that real in-situ ETEM is needed in many TEM characterization analyses, but accommodation of gas or liquid in an electron microscope column is by no means easy. An environmental cell (E-cell) in the microscope column is a key to confine the gas or liquid to the specimen area while the high vacuum level in the remaining parts of the microscope column should not be affected. There are two major types of E-cells: window-type E-cell and differentially pumped E-cell [32]. Designs are totally different and each has advantages and disadvantages.

3.3.2.1 Windows-Type Gas E-Cell

The window-type E-cell concept has a long history back to 1935. Shortly after the invention of the transmission electron microscope, a gaseous environment was introduced through the window-type E-cell to analyze hydrated biological samples and to investigate the contamination rate of the microscope column when using E-cells [32–34]. The so-called window-type E-cell is built in a TEM specimen holder at the tip area, Fig. 3.8. The TEM specimen sits inside the cell, which is then sealed by two windows above and below the specimen. The window materials must be electron transparent and have a weak interference with the electron beam. This means that the window materials should be amorphous in structure and thin enough. The window membranes must also be strong enough to withstand the pressure difference between the gas cell and the TEM vacuum. Typical window materials are amorphous carbon or silicon nitride with a 15–200 nm thickness depending on the desired gas pressure for the applications. The reported maximum gas pressure is 1 atmosphere while TEM imaging and electron diffraction could be done through a 15 nm-thick silicon nitride membrane window E-cell in a 300 kV transmission electron microscope [35]. The cell can be either standalone or with gas inlet and outlet ports.

Fig. 3.8 Scheme of a window-type gas-flow E-cell in an electron microscope column. Gas is contained inside the cell without leaking into the microscope column

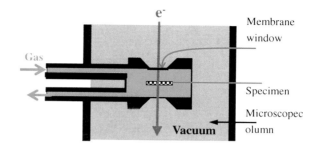

The biggest advantage of the window-type E-cell is the relatively low cost and microscope choice flexibility. Because the E-cell is built into the TEM specimen holder, no modification of the electron microscope is involved (microscope modification is expensive especially when the vacuum system is included) and the holders associated with any electron microscopes can be modified into the E-cell holders. High gas pressure is practical when choosing a robust window design. However, significant disadvantages prevented the window-type E-cell to become popular so far. First of all, it is tedious to assemble an E-cell, and the translation of specimen into the E-cell is not easy, if not impossible. As a consequence, sample exchange is a time consuming challenge. The searchable area on the TEM sample is quite limited. Second, electron scattering from the window materials and the high gas pressure (if applicable) interferes with electron diffraction, deteriorating image quality and resolution. For example, more than 90 % of the electrons at a 300 kV accelerating voltage are lost after penetrating through two 15 nm-thick silicon nitride window membranes and a 1 mm-thick air layer at 1 atmosphere gas pressure. Increasing the accelerating voltage helps to improve image quality, which was why ultrahigh voltage (million volts) transmission electron microscopes were pursued in 1970s for the window-type E-cell applications [32, 36]. However, expensive instruments, high maintenance costs, and high-energy electron beam-caused radiation problems eventually raised serious concerns. The request for ultrahigh voltage environmental transmission electron microscopes went down sharply in 1980s. Third, windows of E-cells may easily get contaminated as a result of deposition of materials evaporated from the specimen by the incident electron beam. Lastly, it is highly risky to heat materials to high temperatures because the window materials may crack, causing gas leakage. It should be noted that progress has been reported recently, where a window-type gas E-cell with a built-in thin wire heater demonstrated a 0.3 nm lattice resolution at 600 °C in a 10^4 Pa pressure environment [35].

3.3.2.2 Differentially Pumped Gas E-Cell

Limited by the obvious disadvantages of the window-type E-cells, attempts were made to leak gas into the microscope specimen chamber directly. In 1965, L.F. Allard and W.C. Biglow explored the possibility by finding out that a vacuum level of $\sim 10^{-2}$ Pa was tolerable in the column of a transmission electron microscope

Fig. 3.9 Schematic illustration of working principle for a gas injection-heating holder. A heater heats specimen while a gas nozzle sprays gas directly onto specimen inside a transmission electron microscope (Patent of Hitachi High Technologies Corporation, Japan)

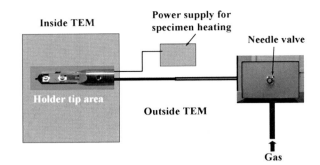

in their lab at the University of Michigan. They then managed to leak air into the microscope column to degrade the vacuum from 4×10^{-3} Pa to 7×10^{-2} Pa while a home-built TEM heating stage was used to heat a single crystalline Cu specimen. Real-time observation of the oxidation process was recorded via a 16-mm cine camera system.

A better method was reported in 1971 by Swann and Tighe [37–39]. An E-cell was built inside a transmission electron microscope, called a differentially pumped E-cell. This was a concept reported by Ruska in 1942 [3]. In this type of design, a significant modification of the electron microscope pumping system is needed. Gas can be injected into the specimen chamber directly either through a port on the pole piece [41–44] or through a specimen holder with a gas injection port and pipeline [27]. Figure 3.9 shows a picture of a gas injection-heating holder reported by Kamino et al. A gas nozzle is placed about 1 mm away from a wire-heating filament. With this design, gas can be sprayed directly onto a TEM specimen under heating [27]. In order to confine the gas within the microscope specimen chamber, a pair of gas restriction apertures should be added along the microscope column above and below the specimen position, Fig. 3.10. Typical bore diameters of these apertures are 0.1–0.3 mm depending on aperture positions. The apertures allow the electron beam to pass through for TEM observation but restrict gas leak from the specimen chamber to the rest of the microscope column. Although small gas leak is still inevitable, the leaked gas is pumped out by the turbo molecular pumps or molecular drag pumps between the restriction apertures [40–47].

The advantages of the differentially pumped E-cell are obvious over the shortages of the window-type E-cells. Because of non-existing window membranes above and below the specimen, the specimen is directly exposed to the electron beam, facilitating high-resolution TEM. For the same reason, heating the specimen to high temperatures is safe and specimen translation is not an issue. The constant gas flux also helps to reduce contamination build-up on the specimen. An apparent disadvantage is the increased cost for microscope construction because of the significant modification of the electron microscope columns. The restriction apertures below the specimen also limit high angle diffraction electrons to be recorded (not good for dark-field imaging). In addition, the maximum gas pressure in a

Fig. 3.10 Schematic illustration of a differentially pumped gas E-cell, which is a built-in part of an environmental transmission electron microscope. Gas is inlet into the specimen chamber and the restriction apertures *above* and *below* the specimen chamber reduce the gas leaking from the specimen chamber to the rest of the microscope column. Extra pumps are necessary to evacuate the gas leaked to *upper* and *lower column*

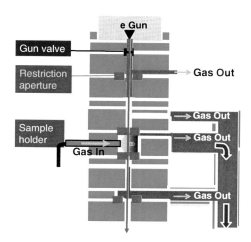

differentially pumped E-cell is limited to below 10^4 Pa in order to maintain the necessary vacuum level in other parts of the electron microscope column. In particular, electron microscopes powered by field emission guns typically require a vacuum of 10^{-8} to 10^{-9} Pa in the gun area, and such a high vacuum is difficult to maintain if the gas pressure is high in the differentially pumped E-cell chamber. In this sense, a LaB_6 electron emitter is better because it works in a vacuum level 3 to 4 orders of magnitude lower than that for a field emitter. The LaB_6 gun transmission electron microscopes have other advantages in the in-situ TEM applications. As the delocalization imaging artifact known from atomic resolution images is much less, therefore, atoms at the sample edges or grain boundaries can be clearly imaged. The high beam intensity of LaB_6 emitters is also ideal for high resolution in-situ TEM imaging. Needless to say, a LaB_6 transmission electron microscope is more affordable in terms of costs of purchasing and maintenance. With the technology advancement, affordable differentially pumped E-cell transmission electron microscopes with LaB_6 emitters became available recently [24, 27, 46].

In-situ gas ETEM has been widely applied to study structural evolution, phase transformation, shape dynamics, crystal or nanocrystal growth, catalysis processes, and oxidation/reduction behaviors [24, 27, 41–45, 48–54]. In addition, catalytic polymerization [55] and gas effect on dislocations [40] were also reported. Figure 3.11 shows an example of Si oxidation studied in a 300 kV transmission electron microscope with a differentially pumped system and a gas injection-heating holder [27]. The as-synthesized Si has a 3 nm-thick amorphous silicon oxide layer on the surface, Fig. 3.11a. The sample was first heated to 700 °C in a vacuum of 3×10^{-5} Pa. An electron beam of ~20 A/cm^2 beam current density was used for real-time, high resolution imaging during heating. The combination of the increased temperature and the focused beam irradiation knocked off the amorphous surface layer as shown in Fig. 3.11b. While the temperature was maintained at 700 °C, air was slowly sprayed onto the Si sample and the pressure

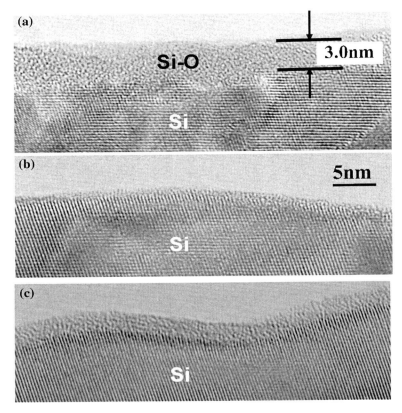

Fig. 3.11 In-situ TEM observation of Si oxidation at 700 °C. **a** As-prepared Si sample with a 3 nm-thick amorphous silicon oxide layer on the surface. **b** Heating the Si sample to 700 °C in a 300 kV H-9500 transmission electron microscope to burn off the surface oxide layer using high electron beam intensity. **c** Re-oxidation of the Si surface in an air pressure of 8×10^{-3} Pa at 700 °C

in the specimen chamber was increased to 8×10^{-3} Pa. Re-oxidation of the Si surface was observed and a 2 nm-thick amorphous surface layer was formed after about one hour as shown in Fig. 3.11c. EELS analysis confirmed the Si and O composition in the newly formed surface layer.

One last thing to address is the gas source used for in-situ gas ETEM. Typical gases used in the study of catalysts and nanomaterials include air [24, 27, 48], O_2, H_2 [40, 42, 49], N_2 [41, 49], He, C_2H_2 [56], C_3H_6 [55], Si_2H_6 [51, 54], Ge_2H_6 [52], and water vapor [43]. The easiest gas source is air, which can be leaked into the TEM specimen area through a built-in pipeline in a specimen holder [27] or a gas port on the pole piece of an electron microscope [40–42]. Home-made gas supply systems are usually simple and affordable, containing a couple of gas bottles and a gas mixture buffer tank. A needle valve or similar is often used to control the gas injection flow rate and consequently the gas pressure in the microscope specimen chamber. If quantitative gas supply is required, the gas supply system needs to be more elegant. It generally contains 4 to 5 types of gases and a water bubbler (to provide water-saturated gases). The gas supply systems containing more than 10 types of gases are rare but exist. For quantitative gas mixture and delivery, high precision gas pressure gauges, mass spectrometers or gas chromatographs should be used. Computer-controlled commercial gas handling systems are available nowadays. Safety is most important, the gases in use should not cause damage to microscope components or injury to users. The room must be ventilated adequately all the time during the use of gas supply.

3.3.2.3 Spherical Aberration (Cs)-Corrected Gas ETEM

In the early years, the window-type E-cell was popular, and a wide pole piece gap was required to accommodate the thick E-cells. The thickness of the E-cell and the wide pole piece gap required ultrahigh voltage transmission electron microscopes (1 MV or higher) to deliver a powerful enough beam able to penetrate and to achieve sufficient image resolution. With the faded ultrahigh voltage microscopes and the advent of the Cs corrector for high resolution TEM [57, 58], the most attractive idea is to build the intermediate voltage (200 or 300 kV) environmental electron microscopes with aberration correctors. The aberration corrector helps to achieve an atomic resolution (0.1–0.2 nm) with a fairly large pole piece gap like 5–10 mm. With the progress in developing combined spherical and chromatic aberration correctors, atomic resolution should be possible to achieve even with a 10–20 mm wide pole piece gap. Such a wide pole piece gap will provide space for insertion of multi irradiation sources or manipulators to the microscope specimen chamber area.

An example of the Cs-corrected gas E-cell transmission electron microscope is the one installed in Brookhaven National Laboratory (USA) in 2007. A built-in differentially pumped system can hold a gas pressure up to 20 mbar (2000 Pa) in the specimen chamber. The pole piece gap is 5.6 mm and a 0.78 Å point-to-point resolution is attainable when gas injection is not in use. The cost for such a microscope is about 5–6 millions US dollars, and therefore is not something affordable by most research organizations. In addition to the high cost, the accessories used on such microscopes often considerably compromise the microscope performance because the aberration-corrected electron microscope is highly sensitive to its working environment. The deteriorated resolution may be due to manipulation controllers, power supply units, cables, the gas supply system, and extra pumps supporting the gas E-cells. More data are required to justify the merit of the very expensive aberration-corrected environmental electron microscopes.

3.3.3 In-situ Liquid ETEM

Both the window-type gas E-cell and the differentially pumped gas E-cell introduced in Sect. 3.2 can be modified to accommodate liquids.

3.3.3.1 Window-Type Liquid E-Cell

Marton was probably the first one to open the field of window-type liquid E-cell [33]. As early as 1935, he attempted to use two 0.5 mm-thick aluminum foils as windows to make liquid E-cells. Abrams and McBain in 1944 [59, 60] used plastic film windows less than 100 nm in thickness, and since then many window-type liquid cells adopted windows made from plastic films, coated with one or more additional layers of evaporated material like silicon monoxide, silicon dioxide, silver, or gold. Reviews for the designs and operation of E-cells in the early years can be found in [32, 36, 61–64].

One of the pioneers for applying the state-of-the-art liquid E-cell TEM in advanced materials science is an IBM group led by F. *Ross*. She and collaborators designed a window-type liquid E-cell TEM specimen holder to study electroplating processes at a nanoscale. In their liquid cell design, two Si wafers were used as the main frame. A $100 \times 100 \ \mu m^2$ window was cut out in the middle of each wafer and covered with an 80 nm-thick SiN membrane. The membranes were thin and transparent to electrons so the electron beam accelerated by a 300 kV TEM could penetrate through. On one Si wafer, a 0.5–1 μm-thick SiO_2 spacer was patterned, a sealed compartment was therefore formed between the two Si wafers when glued together by epoxy. Liquid reservoirs were built to supply liquid electrolyte to the liquid cell and electrodes were made inside and out of the liquid cell. Using such a liquid cell TEM holder, ~5 nm resolution was achieved on an H-9000NAR 300 kV LaB_6 emitter transmission electron microscope. Ross and her colleagues studied the electroplating process of Cu on gold electrodes, the relationship between Cu nucleation rate and electric current density, the diffusion-limited deposition and stripping of Cu, as well as the preferred nucleation sites were discussed [65].

Following exactly the same concept but changing materials and component dimensions, H. Zheng and her colleagues probed the growth kinetics of Pt nanocrystals [66]. The liquid cell they used contained a 200 nm-high liquid compartment filled with 100 nanoliter precursor solution (Pt reagents and surfactants). In a short time intensive illumination under a 300 kV electron beam (LaB_6 emitter), Pt nucleation and growth were triggered. In the recorded streaming video files, mechanisms for Pt nucleation were discovered. Nuclei could grow either by adsorbing monomers from solution or merge with others to evolve into larger single crystalline nanocrystals. Regardless of the growth mechanisms, all of the Pt nanocrystals ended up with a similar shape and size after a certain period of growth time. The same method was then applied to study the real-time growth of Pt_3Fe nanorods at an atomic resolution [67], and to investigate the detailed structural evolution process that happened when the iron oxide nanoparticles in solution approached and attached to each other [68].

Although most high-resolution in-situ applications of the liquid E-cell reported in recent years were based on TEM observations, there exists a chromatic aberration effect for TEM imaging through thick liquid cells. Two approaches can be made to improve the imaging quality at the fixed electron beam energy: one

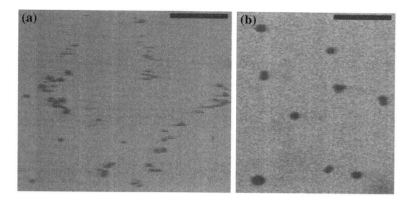

Fig. 3.12 STEM images of gold nanoparticles in water using a liquid E-cell holder. **a** 2–5 nm-diameter gold nanoparticles in water. The observed streaking was caused by particle movement under the influence of the scanning electron beam. **b** 10–20 nm-diameter gold nanoparticles in water. The scale bars correspond to 100 nm (Courtesy of Dr. James Evans, University of California at Davis, and Hummingbird Scientific)

is to make liquid cells and window membranes thinner and also minimize the liquid amount to a level just enough for wetting specimens. The single carbon atomic layer graphene was found to be a perfect support material to encapsulate a liquid solution for high resolution liquid TEM because of its high shape flexibility, mechanical tensile strength, and impermeability to small molecules. In combination with aberration corrected TEM, true atomic resolution was achieved in observation of the growth behavior of colloidal Pt nanocrystals in solution, the motion of as small as 0.1 nm radius Pt nanoparticles could be tracked [69].

Another approach is to consider alternative imaging modes, for example scanning transmission electron microscopy (STEM). Compared with TEM, STEM imaging is less affected by chromatic aberration due to the fact of no lenses below the specimen in the STEM ray diagram. de Jonge et al. achieved a better than 4 nm resolution in the STEM images for liquid materials [70, 71]. In their work, the liquid cell design was similar to the IBM version [65] except flowing liquid in the liquid compartment was enabled. Biological whole cells were grown directly on Si-N membrane windows and labeled with gold nanoparticles. The gold nanoparticles dispersed in the liquid were used to measure the STEM imaging resolution and the result was consistent to the theoretical calculation. In another example, Au nanoparticles with 2–5 nm and 10–20 nm in diameter, respectively, were imaged using a sealed, micro fabricated fluidic cell holder, Fig. 3.12.

It seems not impossible to obtain a better than 1 nm STEM resolution for imaging materials in liquid, but achieving much higher than 1 nm STEM imaging resolution may be obstructed by Brownian motion in liquid.

Strictly speaking, there is also the third way to get around the considerable chromatic aberration problem caused by the window-type liquid E-cell. The method is to accommodate liquid around the specimen without using the window-type liquid cells.

3.3.3.2 Liquid ETEM with an In-Column Differentially Pumped Liquid E-Cell

Several types of differentially pumped liquid E-cells were developed since 1968 in order to carry out electron microscopy studies on live cells and get electron diffraction patterns from wet biological crystals and wet cell membranes [72, 73]. The big advantages of accommodating liquids directly in electron microscopes are again the high resolution and lowered chromatic aberration effect because the total sample thickness is reduced. Gai et al. developed a TEM specimen holder to inject microliter liquid into the electron microscope specimen chamber and the specimen could be heated to elevated temperatures [74]. Using this holder on a modified 200 kV transmission electron microscope with a differentially pumped system, heterogeneous hydrogenation of polymerization in the presence of liquid, catalysts, hydrogen gas, and a thermal field were studied.

Another intriguing discovery that may lead to a progress in accurate delivery of small quantities of liquid into the local area on a TEM sample was reported by Sutter and Sutter [75]. In this work, a pipette made from a Ge nanowire as a body and an Au-Ge reservoir wrapped by a carbon shell as a liquid source were demonstrated. When heating to higher than 361 °C (eucentric temperature of bulk Au-Ge alloy), the Au-Ge alloy melted but was confined within the carbon shell. Once a fine hole was pierced into the carbon shell using a focused electron beam in a 300 kV transmission electron microscope, the liquid leaked out and formed dispensed liquid drops 20–40 nm in diameter.

3.3.4 In-situ Biasing TEM

In-situ biasing TEM, by its definition, means TEM observation of a specimen on which a bias voltage is applied. For electrical conductive specimens, a circuit can be formed with a TEM specimen as a component for electric current to pass through. In-situ biasing TEM is often used to study behaviors and structural changes of specimens in an electrical field, as well as electrical transport properties of individual nanostructures and Joule heating effects in microscale and nanoscale structures. Probe TEM specimen holders are popular for these application purposes. A built-in metallic probe with a tip of 50–100 nm in diameter can move back and forth in fine steps controlled by a piezo manipulator, so that gentle approach and contact is possible between the probe and the nanostructures. The probe is connected to a power supply unit by which a voltage can be applied to the specimen, which is at grounding potential. The probe can also be used to manipulate nanomaterials. All these are carried out with simultaneous TEM observations.

Using a probe TEM holder, a single-walled carbon nanotube of 12 nm in diameter and 24 nm in length was stretched in a transmission electron microscope to study the plasticity [76]. A constant bias of 2.3 V was applied and the Joule heating caused

Fig. 3.13 Four TEM images, spaced by 1-min time increments, showing indium transport on a multi-walled carbon nanotube. The tip closer to the anode which is out of view to the *left* has a higher temperature than the tip to the cathode (out of view to the *right*). The mass transport is driven by the temperature gradient (Courtesy of Zettl Research Group, Lawrence Berkeley National Laboratory and University of California at Berkeley)

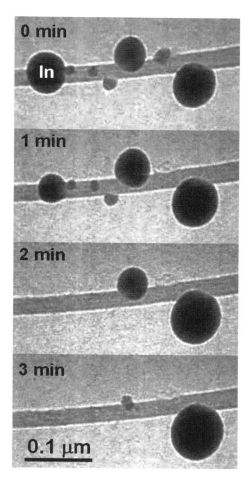

a temperature rise to about 2000 °C in the carbon nanotube. A 280 % tensile strain with a 15-fold reduction in diameter seen in TEM images revealed the superplasticity of this nanotube although a possible role of electron beam irradiation effect was not addressed. Using the same in-situ biasing TEM holder and experimental parameters, exceptional plasticity of double-walled and triple-walled carbon nanotubes (190 % elongation and 90 % diameter reduction) [77], and amorphous-to-crystalline structural transformation of carbon nanowires were also reported [78].

In another excellent example, nanoscale mass transport along carbon nanotubes was realized [79], Fig. 3.13. First, indium metal was evaporated onto multi-walled carbon nanotubes, which formed nanoparticles attached to the walls. Using a probe TEM holder, electric current was driven through the carbon nanotubes and the Joule heating raised the local temperature, resulting in a temperature gradient along the longitudinal axis of the nanotubes. When the local temperature was above the melting point of the indium particles, atomic scale mass transport of indium to cooler zone was observed. The mass transport was reversible when the direction of the current flow was opposite [79].

Using the same principle, nanoscale cargos that not only translated along but rotated around a multi-walled carbon nanotube were observed using in-situ scanning electron microscopy (SEM) [80]. Although not mentioned clearly, the fabrication of the co-axial carbon nanotubes and application in the nanoscale cargo transport and rotation experiments was similar to earlier results reported by Cumings et al., in which a probe TEM holder with four electrical leads was used to open the tip of a multi-walled carbon nanotube and to peel off the tube wall layers [81] or pull the inner tubes out of the outer shells [82], Fig. 3.14.

A bias voltage of 2.9 V generated a 200 mA electric current through the multi-walled carbon nanotube and burned off the enclosed tip. Some outer tube walls were peeled away so the remaining tip became much sharper. The successive peeling and sharpening dynamic process were observed in a high-resolution transmission electron microscope [81]. The authors further demonstrated that the exposed inner tubes could be pulled out of the outer walls and then pushed back using the probe as a manipulator [82]. The TEM images and corresponding schematic representation in Fig. 3.14 clearly show the inner tubes sliding out and in, similar to an old style telescope. The result indicated an ultralow friction between each concentric tube walls and the multi-walled carbon nanotubes hold a great promise for nanomechanical applications because of the low-friction and the low wear nanobearing property [82].

As noticed a couple of times already, Joule heating is often an important part of the experiments using an in-situ probe TEM holder (e.g. Fig. 3.13). It can of course also be used for in-situ heating TEM observations. For example, electrically driven redox processes in cerium oxide films were studied. Ordered oxygen vacancies could be generated in CeO_2 thin films by passing an electric current (6 V) through the films. The films were oxidized again in the microscope chamber when the current was turned off [83].

For in-situ electrical transport measurements in transmission electron microscopes, the key is to make an ohmic contact between the nanostructure and electrodes provided by the in-situ probe TEM holder. In a successful attempt, conductance of an individual carbon nanotube was measured using a probe TEM holder, which was correlated to the structural changes imaged simultaneously [84]. In this work, a carbon nanotube with one end fixed to a conductive probe manipulator was pushed into a liquid metal (Hg or Ga) to make Ohmic contacts between the liquid metal and the suspending ends of the nanotubes. Two-point current-voltage (I-V) measurements were done while doing in-situ TEM observation. A ballistic transport feature at room temperature was revealed [85]. It was also demonstrated in other in-situ biasing TEM experiments that structural deformation plays an important role in the electrical conductance of carbon nanotubes [85, 86]. Similarly, electrical resistivity of ZnO nanowires was found to be linearly proportional to the bending degree [87].

In addition to the properties of individual nanotubes or nanowires, interconnections among nanostructures are also of great interest. Various shapes

Fig. 3.14 A series of
TEM images showing
telescoping of a multi-walled
carbon nanotube using an
in-situ probe TEM holder.
The model at the *bottom*
illustrates the sliding of the
inner tubes with respect to
the outer shells as observed
experimentally (Courtesy of
Dr. John Cumings, University
of Maryland)

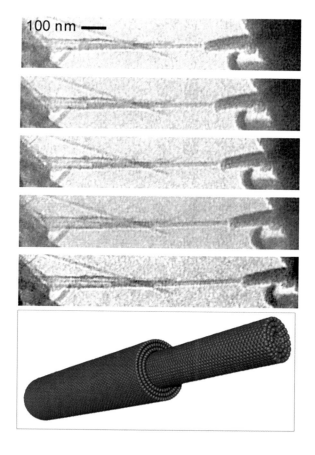

of connections between carbon nanotubes were successfully fabricated taking
advantages of a precise positioning control (0.02 nm) and in-situ soldering by electron beam-induced amorphous carbon or Joule heating-induced tungsten melting
soldering. I-V curves of the tube/tube connections were measured and the junction
resistance was lowered by graphitization of amorphous carbon bonding material
[85–88]. Tips of carbon nanotubes were also modified in a transmission electron
microscope and the I-V measurements clearly demonstrated the importance of the
tip structure to the field emission characteristics [88]. The electric field not only
triggered a field emission from the tips of carbon nanotubes, but was also used to
field evaporate carbon nanotubes segment by segment, and the electron beam was
intentionally used to adjust the field evaporation rate in a nanoscale area to modify
the tip morphology [89].

In-situ biasing TEM is certainly useful in the study of nanoscale films and
electronic devices. Local electrical transport properties in nanoscale magnetic
tunnel junction thin films [90] and the effects of annealing on electrical transport of magnetic tunnel junctions [91] were investigated. In particular, tunneling

Fig. 3.15 A picture of the tip area of a dedicated electrode TEM holder. Voltages of up to 1 kV can be applied to the specimen (Patent of Hitachi Materials Research Lab, Hitachi Ltd)

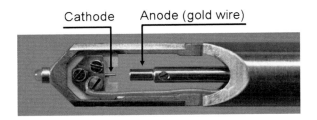

magnetoresistance (TMR) of individual nanomagnetic tunnel junctions under working conditions were measured using the in-situ probe TEM holder [92]. Because the nanomagnetic devices were very small (100×150 nm^2 in area), in-situ TEM observation was essentially the only way to make sure that the piezo controlled probe could make a correct contact on the single device to be measured.

With great attention paid to the energy-related materials and processes in recent years, in-situ biasing TEM was recently employed to study the real-time charge-discharge process in nanomaterials-based lithium-ion batteries. Huang et al. built a nanoscale electrochemical device in an in-situ probe TEM holder and successfully observed the structural and morphological changes of the SnO_2 nanowire anode during the lithiation process. The SnO_2 single crystalline nanowire was converted to a Li_2O amorphous nanowire containing embedded Sn and Li_xSn nanoparticles and the process resulted in more than 200 % volume dilation [93].

Compared with the in-situ probe TEM holders, the dedicated electrode TEM holder has a simpler design as shown in Fig. 3.15 but a much higher voltage can be applied.

Anode (gold wire) and cathode are integrated into the tip of a TEM holder. The distance between anode and cathode can be adjusted from 0 to 5 mm, and voltages of up to 1 kV can be applied [94, 95]. The specimen to be studied is mounted on the cathode and the distance between the specimen and the anode is typically 0.5 mm. A voltage applied via the cathode causes a potential difference across the specimen. Such a holder has been used together with the in-situ Lorentz TEM to investigate the field emission of multi-walled carbon nanotubes [94, 95], and high voltages (800–900 V) were found particularly valuable in these studies [96]. Also, using a specially designed field-effect transistor-based biasing holder, chiral indices and chirality-dependent transport properties of double-walled carbon nanotubes have been studied [97].

3.3.5 In-situ Nanomechanical TEM

Mechanical properties such as strength, hardness, and toughness may be changed significantly when the sizes of materials go down to nanometer scales. With increasingly miniaturized electronic components and the request to correctly understand and predict failure of nanostructures, material size effects on mechanical

Fig. 3.16 TEM image of a
CdS spherical nanoparticle
under a compression
force. The corresponding
Displacement-Force curve
acquisition and TEM
observation were performed
simultaneously in real time.
The sudden load drop at
about 6 mN was not due to
a catastrophic failure but
sliding of the nanoparticle
away from the site as
revealed by TEM observation
(Courtesy of Dr. Zhiwei Shan
of Hysitron Inc. and Dr. Andy
Minor of Lawrence Berkeley
National Laboratory)

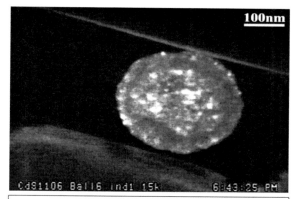

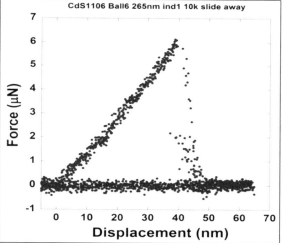

behaviors become more important and urgent than before. Like all other in-situ TEM technologies, the development of in-situ nanomechanical TEM was because of the high resolution that TEM can offer for direct imaging and characterization of internal structures and defects in materials. A good example was given by Z.W. Shan et al. for why in-situ TEM is necessary for nanomechanical testing [98].

Figure 3.16 shows a CdS spherical nanoparticle under a constant compression force. The ball was gradually compressed and the corresponding Displacement-Force curve tells a linear relationship between force load and deformation displacement. The load drops suddenly at about 6 μN, which may imply a catastrophic failure of the ball. However, simultaneous TEM observation told a different story, the CdS nanoparticle did not break but slid away. This beautiful "seeing is believing" example again reminds a possibility of misinterpretation of experimental data if real-time in-situ TEM observation is not available.

In fact, in-situ experiments in the late 1960s and early 1970s already showed a capability for direct observation of mechanical indentation, compression, and bending deformations in scanning electron microscopes and transmission electron microscopes [99, 100]. After decades of development, precisely controlled, quantitative in-situ nanomechanical TEM holders have been developed to characterize the mechanical behaviors and corresponding structural changes in micro and nanostructures [101]. One of the pioneering works on the in-situ nanomechanical TEM was to measure the elastic modulus of carbon nanotubes [102]. In this work, a custom-built TEM holder was used in which carbon nanotubes were attached to a gold wire through which an electric potential was applied to the nanotube. Static and dynamic mechanical deflections were induced in the multi-walled carbon nanotube cantilevers in a transmission electron microscope [102]. Although the elastic properties were successfully derived in the experiments, the method used is not easy to be expanded to nanomechanical testing of other nanostructures of various shapes. Obviously, specially designed straining TEM holders are required to load a tensile or compressive mechanical force on TEM specimens while tracking real-time structural changes in order to find out and quantify mechanical property-structure relationships. Various TEM holders with different designs have been reported, including conventional tensile straining holders, microelectromechanical systems (*MEMS*)-*based straining* holders, nanoindentation holders, and TEM grid-based straining holders.

3.3.5.1 Conventional Tensile Straining TEM Holder

The concept of the conventional straining holder for in-situ electron microscopy originated in the late 1960s [103–105]. The basic idea was to make a specimen in a rectangular shape with one end fixed and another end connected to a movable rod, which could slide along the length of the holder. Figure 3.17a shows a picture of such a holder (tip part) for in-situ tensile straining TEM. The letter p labels two pins, the pin to the left is connected to a sliding rod. The sample is made into an 11×3 mm rectangular shape as shown in Fig. 3.17b. The central area is thinned for TEM observation and the two holes on both sides of the thinned central area are used to positioning the sample along with the pins shown in (a). Clamps and screws are used to secure the sample into the holder tip. When pulling the sliding rod, a tensile straining force is applied to the central sample region where in-situ TEM observation is performed [106]. The sample shown on the right side of Fig. 3.17b is after in-situ straining TEM.

As an application example, Fig. 3.18 shows a real-time TEM observation of dislocation-precipitate interaction processes in a stainless steel sample under a tensile straining force. The dislocations pinned by carbide precipitates eventually bypassed the pinning sites via either a simple mechanism or a complex configuration [107]. In particular, roles of hydrogen in enhancing the mobility of the dislocations and in determining the cracking rate along certain directions were discovered using a tensile straining holder working with a gas E-cell [41, 107]. Using a similar specimen holder, dislocation dynamics, as well as fracture and plasticity deformation mechanisms in nanocrystalline nickel were studied comprehensively [108–110]. Heating or cooling is also possible for some straining TEM holders.

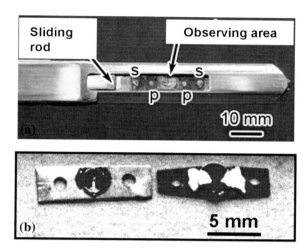

Fig. 3.17 **a** A picture for the tip area of a conventional tensile straining holder. **b** A specimen (*left side*) prepared for in-situ tensile straining TEM, two holes at the two ends are used to align the specimen with the two mounting pins on the holder (labeled *p* in **a**). The pin to the *left* in **a** is connected to a movable rod. Clamps are placed over the mounting pins and the specimen is further secured to the position via screws (labeled s in **a**). The specimen to the right in **b** is after a straining TEM experiment (Courtesy of Dr. Ian Robertson of University of Illinois at Urbana Champaign)

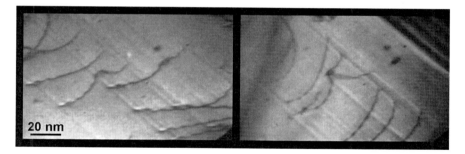

Fig. 3.18 In-situ TEM images showing dislocation-precipitate interactions in a stainless steel specimen at room temperature. The dislocations pinned by precipitates (*black dots*) could bypass the pinning sites either in a simpler way (*left*) or in a complex way (*right*) (Courtesy of Dr. Ian Robertson of University of Illinois at Urbana Champaign)

3.3.5.2 MEMS-Based Straining TEM Holder

A deficiency of the conventional tensile straining TEM holder described above is its inability to quantify the force load and material deformation displacement. In addition, only tensile force can be applied. To overcome these problems, MEMS-based mechanical testing holders were designed [106, 111]. The key part, a free-standing MEMS chip device integrated into the tip of the holder, is fabricated using the standard silicon-based microfabrication followed by lift-out from the substrate, Fig. 3.19.

Fig. 3.19 Schematic illustration of a free-standing 10 mm × 3 mm MEMS chip device which can be integrated into the tip of a TEM holder for uni-axial tensile testing

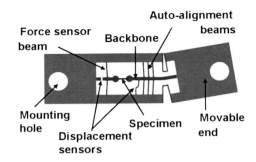

The force load is applied with a displacement-controlled mechanical, thermal, or electrostatic mechanism and the applied load and straining are measureable by displacement-force sensors or differential capacitive sensors. Both tensile and compressive forces are possible to be applied. The dimension and shape of such a MEMS chip device is similar to those of the specimen shown in Fig. 3.17b. Using a MEMS device with thermal actuation and capacitive load sensing, in-situ mechanical testing and TEM observation were performed on multi-walled carbon nanotubes, whereby elastic modulus and fracture strength close to theoretical predictions were confirmed [112, 113]. Details in designs, working principles, experimental examples, and related references for various MEMS devices for in-situ nanomechanical TEM can be found in [111].

3.3.5.3 Nanoindentation TEM Holder

The development of nanoindentation TEM holders catalyzed quantitative in-situ compressive straining TEM studies. Loading rate control at micronewtons/s and displacement rate control at nanometers/s are feasible now driven by mechanical/piezo positioning elements or nanoscale positioning motors. A flat-tip of the nanoindentation holder is made from diamond or tungsten, which is used to punch into a specimen. As can be seen in Fig. 3.20, the indentation is from a side direction, while the electron beam coming from top, images morphological and structural changes in the deforming area. Specimen used with this type of nanoindentation holder are either thin films coated on substrates, nanomaterials attached to supporting subtracts, or crystals made into a pillar shape with a sub micrometer diameter. A focused ion beam (FIB) system is often required to fabricate micro pillars because FIB makes sample preparation site-specific and dimension/geometry-controllable. However, because FIB is Ga ion milling-based, it inevitably changes surface structure of materials and in turn has effects on mechanical testing results. The influence of ion beam milling is increasingly prominent with decreasing object size as a result of the larger surface to volume ratio. Caution is therefore called for data interpretation if specimens are prepared using FIB.

Various nanostructures have been studied using the in-situ nanoindentation TEM holders. A classic in-situ nanomechanical TEM work is shown in Fig. 3.21.

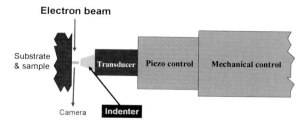

Fig. 3.20 Schematic illustration of a nanoindentation holder for in-situ nanomechanical TEM (Courtesy of Dr. Zhiwei Shan and Dr. Oden Warren of Hysitron Inc.)

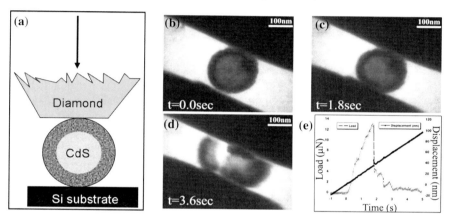

Fig. 3.21 In-situ nanomechanical TEM result obtained using a nanoindentation TEM holder. The specimen was a hollowed CdS nanoparticle. **a** Scheme of the indenting area. For the simultaneous TEM observation, electron beam is parallel to the paper-normal direction. **b–d** A series of TEM images showing morphological changes of the nanoparticle from beginning (no load) to failure at the time of 3.6 s. **e** Time-Load-Displacement curves recorded simultaneously with TEM observation (Courtesy of Dr. Zhiwei Shan of Hysitron Inc.)

A nanocrystalline CdS hollowed sphere with a diameter of 210 nm was applied a compression force in a transmission electron microscope. The compression rate was 20 nm/s and Time-Load-Displacement curves were recorded in real-time accompanied by the correlated shape change revealed by time-resolved TEM imaging [114]. Figure 3.21a shows the beginning state and Fig. 3.21b corresponds to the peak load of 13.2 μN right before fracture is observed in Fig. 3.21c. An effective stress of 370 MPa, a compressive strain of 16.4 %, and a stiffness of 2.3 GPa were derived. The values indicated that the stress in the nanocrystalline hollow sphere was quite large although the effective stiffness was much smaller than in the bulk counterpart [114].

In-situ TEM mechanical testing of WS_2 nanotubes was reported using a holder with a sharp W tip as the indenter and a flat Pt tip as a support to WS_2 nanotubes. Both W and Pt tips also served as electrodes to pass electric current through the nanowires [115].

In-situ nanomechanical TEM on sub-micrometer diameter, FIB-prepared, pillar samples has a great advantage in probing the size effect, e.g., uni-axial size-dependent strength. Experiments showed a surprising "mechanical annealing" effect in which defects in single crystalline Ni pillars, 150–400 nm in diameter, could be completely removed simply by applying a mechanical stress [114]. Using the pillar samples, systematic in-situ TEM investigations of size effects were also performed on single crystalline Ti-Al alloy pillars with diameters between 8 and 0.4 µm. 1 µm diameter was found to be a mechanism transition threshold below which the compression deformation mechanism changed from deformation twinning to dislocation plasticity. The maximum flow stress saturated at a value close to the ideal strength of bulk Ti [116].

3.3.5.4 TEM Grid-Based Straining Holder

Recently, an innovative method was developed to conduct in-situ tensile and bending of nanowires in a transmission electron microscope [117–122]. An epoxy with desired strength was used to make colloidal thin films (polymer with high strength and transparency) supported by commercially available 3 mm-diameter TEM grids. The thin films were intentionally broken prior to or during TEM observation. Figure 3.22a is a schematic illustration of a Cu grid covered with a colloidal thin film on which nanowires were randomly scattered. In a transmission electron microscope, the colloidal thin film was heated or irradiated by electron beam to make it curling or shrinking. Both mechanisms usually started from the broken edges as shown in Fig. 3.22b and c. As a result, some nanowires on the film were wrapped into the curled thin films, they were either bent [120, 121] or under an axial tensile force [117–119].

Figure 3.23 shows in-situ TEM images of an amorphous silica nanowire. A uniform and dramatically reduced wire diameter accompanied by axial elongation indicates a super-plastic mechanical property of this specific nanowire [119]. The initial nanowire diameter is 36 nm, which decreases to 20.3 nm after tensile deformation. The deformation spreads across a rather wide region (~700 nm, marked by the two white bars in Fig. 3.23h) without obvious necking (an indicator of quick glass break). This tensile experiment indicated a super-plasticity of about 215 % for the silica nanowire studied.

Using the same method, in-situ tensile straining TEM investigations were performed on Si nanowires [117], and SiC nanowires [118]. Figure 3.24 shows a case of bending a Si nanowire. The bending was imaged in real time imaged at low and high magnifications. The high-resolution TEM images show full dislocation activities (c and d) [121]. Distinctive features of the incipient elastic-to-plastic transitions between the tensile and compressive regions of the bent nanowire are clearly seen. Burgers vectors of the dislocations are determined from the high-resolution TEM images (c and d) [121]. Bending of SiC nanowires was also investigated, elastic deformation, elastic-to-plastic transition, and plastic deformation processes were successfully captured at nanoscale and atomic scale in a transmission electron microscope [122, 123].

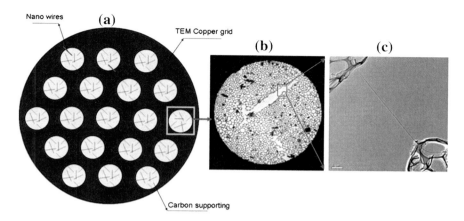

Fig. 3.22 The colloidal thin film contraction method for generating a tension force on nanowires. **a** A drawing of a TEM Cu grid. **b** TEM micrograph for a hole on a TEM grid covered with a carbon supporting film. **c** Pulling of a single nanowire by the contracted carbon film (Courtesy of Dr. Xiaodong Han and Dr. Ze Zhang of Beijing University of Technology, China)

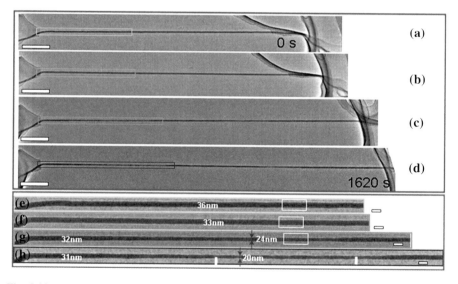

Fig. 3.23 In-situ TEM images showing a superplastic elongation of a silica nanowire under a tensile force and a moderate electron beam illumination. **a–d** The large strains, with the framed areas magnified in **e–h**. Scale bars **a–d** 500 nm, **e–h** 50 nm (Courtesy of Dr. Xiaodong Han and Dr. Ze Zhang of Beijing University of Technology, China)

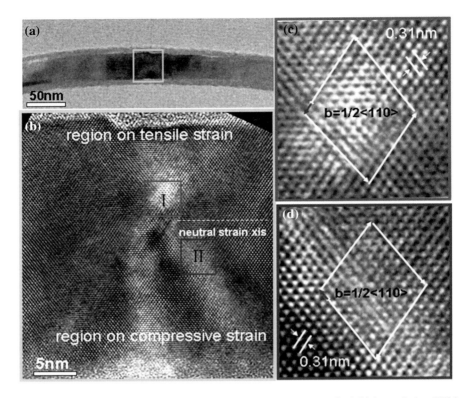

Fig. 3.24 A bent Si nanowire. **a** A low magnification TEM image. **b** A high resolution TEM image taken from the framed region in **a**. **c** and **d** show high resolution TEM images of the framed regions *I* and *II* in **b**, respectively (Courtesy of Dr. Xiaodong Han and Dr. Ze Zhang of Beijing University of Technology, China)

This section about in-situ nanomechanical TEM focuses on typical instruments, methodologies, and material systems. Nevertheless, nanomechanical property testing is not limited by the aforementioned in-situ nanomechanical TEM holders or methods. For example, it was reported that multi-walled carbon nanotubes could be turned into a pressure cell to study the effects of high pressure on nanomaterials [124]. In this example, carbon nanotubes filled with Fe_3C, Fe, or Co were heated to 600 °C in a 300 kV field emission transmission electron microscope. The purpose of the heating was to mobilize defects in the graphite structure of the carbon nanotubes so to prevent defect agglomeration, which might destroy the nanotube structures rapidly. Self-compression of the carbon nanotubes induced by electron beam irradiation generated a compression force on encapsulated nanomaterials. ~40 GPa pressure or higher pressure deformed, extruded, and broke hard materials such as Fe_3C as clearly confirmed in the in-situ TEM experiments [124]. Because of the page limit of this chapter, some other areas such as in-situ TEM nanotribology and in-situ mechanical testing on biological materials are not covered.

3.3.6 In-situ Lorentz TEM and In-situ Electron Holography for Imaging Magnetic or Electric Field Distribution

For TEM imaging of the materials containing magnetic or electric field, two TEM techniques are commonly used, Lorentz TEM and electron holography.

3.3.6.1 Lorentz TEM: Principle

This is a phase contrast TEM imaging technique, which is used to observe magnetic domains. It is based on a phase shift in incident electron waves caused by magnetic specimens. Two imaging modes are routinely used. Fresnel mode (under- or over-focus imaging) is used for imaging magnetic domain walls. It is also sensitive to the change of electrostatic potential in specimens. Foucault mode (similar to dark-field imaging) is mainly used to image magnetic domains [125].

In the Fresnel mode, defocus leaves alternating bright and dark lines in a TEM image, the bright lines occur at the position of magnetic domain walls at which the magnetizations on both sides deflect the electrons towards the wall. If a coherent electron source is used, the convergent image of a magnetic domain wall is composed of a set of electron diffraction fringes running parallel to the wall. The over-focus imaging condition causes an image contrast reversal, see example in [126]. To image magnetic domains using the Foucault mode (dark-field-like), the objective lens of transmission electron microscope is kept in-focus, an electron diffraction pattern is first obtained in which diffraction spots split due to the magnetic field in materials. One of the diffraction spot in the split spot pairs is chosen to pass the objective aperture to form microscopy images.

The advantage of the Lorentz microscopy is to directly observe important features such as location and direction of the magnetic domain walls over a large field of view (up to tens of microns) and almost any modern transmission electron microscopes can do the job. The shortage is that the specimen surrounding area in the electromagnetic objective lens must have negligible external magnetic field strength. A popular solution to reduce the influence of the magnetic field coming from the objective lens of electron microscopes is to turn off the objective lens as well as the pre-field objective lens used for focusing the electron beam. The remnant magnetic field at the specimen location can be reduced to 200–300 Oe, which is, however, still influential for some soft magnetic materials. In addition, image focusing will have to be realized using the first intermediate lens and TEM magnification is limited to the use of the second intermediate lens and projector lenses. As a consequence, the magnification power is rather weak, usually lower than $10,000\times$ [127].

The compromised Lorenz TEM performance on standard electron microscopes makes the dedicated Lorentz transmission electron microscope a better choice. The dedicated Lorentz microscope uses a weakly-excited long-focal length objective lens (on the order of more than 10 mm compared to ~2 mm for a standard lens) and the specimen is located above the pole piece gap where the magnetic field is a few thousand times lower than that within the pole piece gap. Recently, a transmission electron microscope with dual specimen stages was invented with a quickly switchable conventional TEM mode and a dedicated Lorentz TEM mode [128–132] therefore the microscope can be used for both routine TEM characterization and for dedicated high resolution Lorentz TEM.

3.3.6.2 Electron Holography: Principle

For materials containing magnetic or electrostatic fields, the phase of incident electron waves are modified when the electron waves exit the TEM specimens. Recovering the phase of the exited electron waves will reveal the electric or magnetic information in specimens. Electron holography is a technology that measures both phase and amplitude of the electron waves. Two or more electron waves with different phases are brought into interference and form interference image (electron hologram), which contains three-dimensional amplitude and phase information at an atomic resolution level. In contrast, the phase information is not available from all other electron microscopy imaging techniques. Off-axis electron holography is the most widely used holography mode nowadays [133], in which an electron wave passes through a specimen (the objective wave) and a reference electron wave travels in vacuum. An electrostatic biprism system below the objective lens combines the objective wave with the reference wave, the interference between the two waves with different phases forms an electron hologram [134–136]. The phase difference stems from a phase shift occurred in the objective wave when passing through the specimen, and can be recovered to retrieve the magnetic or electrostatic potentials in materials. Recent advancement in multiple-biprism electron holography made this technology even more flexible and powerful [128–132, 137–139].

3.3.6.3 In-situ Lorentz TEM and In-situ Electron Holography

If in a TEM study the specimen is under a versatile environment such as temperature change [129, 140, 141], external magnetic field [142–145], bias voltage in active functional devices [146–152], or electric current [139, 153–155], one most likely should consider in-situ Lorentz TEM or in-situ electron holography. If an external magnetic field is required, in-situ TEM holders should be integrated with Helmholtz coils or electromagnets to generate a magnetic field at the TEM specimen area [143, 144, 156–159]. Another method to apply a magnetic field to the specimen inside a transmission electron microscope is to tilt the specimen in the intrinsic magnetic field of the objective lens [160–162]. It is also possible to bring a piezo-driven magnetic needle close to the specimen [163, 164]. The magnetic field in the specimen chamber of a transmission electron microscope can be measured using a modified TEM holder with a Hall sensor, Hall probe, or samples with known magnetic response together with magnetometers [165]

Fig. 3.25 Electron
holograms (*a*)–(*c*), and
corresponding electron wave
phase images (*A*)–(*C*) for
a carbon nanotube applied
with a 0, 70, and 120 V
bias voltage, respectively.
In this field emission study,
the phase images show the
electric field distributions
around the tip of the
nanotube (Courtesy of Dr.
John Cumings of University
of Maryland)

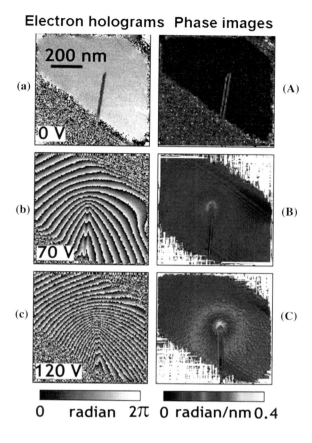

As an example, in-situ electron holography was used to map the electric field distribution surrounding charged carbon nanotubes [166]. The experiments were special because the biprism used for electron holography was also made from a multi-walled carbon nanotube [167]. It was found that a 70 V electric field activated electron emission from the tip of a carbon nanotube. This field emission became stronger at 120 V. Electron holograms were recorded as shown in Fig. 3.25a–c, in which phase shifts of the electron wave at applied bias voltages of 0, 70, and 120 V were revealed as shown in Fig. 3.25a–c.

It can be seen that no electric field exists at the zero bias voltage. At 70 and 120 V bias voltages, high phase gradients, which correspond to electric potentials are seen at the tip of the carbon nanotube specimen. No high concentration of electric potential was seen at other locations along the nanotube length. The electric field strengths were derived based on a comparison between the experimental data and theoretical calculations. The measured data also concluded that for the same bias voltage, the local electric field strength at the tip of the nanotube was inversely proportional to the nanotube diameter [166].

Following the in-situ Lorentz TEM imaging principle, field emission of multi-walled carbon nanotubes were also studied using an in-situ biasing TEM holder (see Fig. 3.15) which allows a bias voltage of up to 1 kV and wide gap of up to 5 mm between the tip of a nanotube and the surface of anode. The wide gap, typically 0.5 mm in the field emission study, helps to avoid the electric discharge and the influence of outgas from the electrode. The large electric field induced high electron emission from the tips of the nanotube specimens and the observed emission current fluctuation at >20 mA was attributed to a peel-off damage of the carbon nanotubes as observed in TEM images [94, 95], and the liquid characteristic of Fe particles at the tips of carbon nanotubes was found to be related to the temperature effect due to the high emission current [96].

3.3.7 In-situ Ion Beam Irradiation TEM

All of the in-situ TEM technologies except in-situ ETEM introduced so far are on the basis of design or modification of TEM specimen holders. Differentially pumped E-cell needs the addition of gas restriction apertures and an extra pumping capability on the electron microscope column. There is another type of technology, in-situ ion irradiation TEM, which also requires extensive modifications on the microscope column. An ion beam is used to bombard the specimen while the effects are monitored in real time using TEM. The bombardment effects include phase changes and segregation, mechanical and structural changes, atomic mixing and chemical disorders, compositional changes, chemical reactions, grain growth/shrinkage, precipitation/dissolution, defect or bubble formation/growth/ motion/coalescence/removal, destruction, ionization, diffusion, and collision cascades. Ion beam irradiation TEM became possible since 1961 [168] and has been well developed and applied for nuclear materials, space and astronautic engineering materials, semiconductor materials, materials for irradiation environments, ion implantation, and deformation mechanisms in materials.

More than 30 units of ion irradiation transmission electron microscopes were ever built in the world since 1961, and 11 of them were still in operation as of 2009 [169]. In most cases, one or two ion beam lines were introduced into the microscope column and guided to the specimen area for bombardment. Ion sources could be light or heavy ions such as H, He, O, Ne, Ar, Fe, Ga, Kr, Xe, or U, and beam energies range from a thousand volts to a few million volts. Usually one ion beam line is not so difficult to interface with an electron microscope at low elevation angles. Adding two or three ion beam lines, especially at high elevation angles requires complicated modifications on the electron microscope column, but the reward is the capability of using more than one irradiation condition to explore the combined effects.

There are many technical requirements for a transmission electron microscope to adopt ion beam lines while it maintains a sound TEM performance. First of all, ports are needed and must be vacuum-tight when connecting to ion beam lines.

Fig. 3.26 300 kV
H-9000NAR transmission
electron microscope with
an ion beam line introduced
from a side direction as
marked in the figure. The
ion beam and electron beam
hit the TEM specimen
simultaneously. This
microscope at Argonne
National Laboratory is the
only one of its kind in the
USA

The pole piece gap should be wide enough to accommodate the beam lines and the associated accessories such as Faraday cups and electrical connections for measuring the ion flux arriving at the specimen. Both extra ports and a wider pole piece gap compromise the TEM imaging resolution. The resolution deterioration may be 30–50 % but atomic resolution should still be achievable in modern high-resolution transmission electron microscopes. The beam lines can target the specimen from high (as high as 90°) or low (as low as 0°) elevation angles but there are some TEM operation limits for the low angle ion beam lines [169]. Active and passive anti-vibration measures on electron microscopes and appropriate room designs are also important to minimize the excessive vibration caused by ion beam lines and ion sources.

One of the best known, active and productive ion irradiation TEM instrument is at Argonne National Laboratory, USA, Fig. 3.26.

It is usually fully occupied by many ongoing materials research projects. One of the applications was to use various heavy ions at different energies to bombard the $YBa_2Cu_3O_{7-d}$ superconductor single crystals to investigate irradiation damages at high resolution [170].

Another case in which a significant modification of the electron microscope is necessary is so-called the ultrafast TEM or dynamic TEM (DTEM). This is a technology that uses two synchronized laser beams with nanosecond or femtosecond pulses to generate electrons for TEM imaging and for inducing structural changes in the specimen. Details are introduced in Chap. 4 of this book and therefore will be skipped here.

3.4 Some Important Notes

3.4.1 In-situ TEM: Holders and Atomic Resolution

From the above introduction into in-situ TEM technologies, one should be able to make an important point: *To some extent,* in-situ *TEM is all about specimen holders.* External fields, forces, or exotic environments can be applied to a TEM specimen area through specially designed specimen holders. There is really not much technical limit to the holder-based in-situ TEM technology if one has great imagination and creativity. Exceptions do exist; they are chamber-based technologies like differentially pumped E-cell (Sect. 3.3.2.2) and ion beam TEM (Sect. 3.3.7).

Taking the most popular in-situ heating TEM as the first example, heating mechanisms vary for different advantages (Sect. 3.3.1). If more than just heating is desired, such as in-situ evaporation deposition, one may squeeze two or three heaters into the holder's specimen cup (Sect. 3.3.1). The second example is about site-specific in-situ heating TEM on interfaces or grain boundaries. Manual cross-sectional sample preparation is not only time consuming but also ends up with random chances of having the interesting, sub-micrometer structural features involved in the TEM observable area. A FIB system is known to be precise for lifting out a microscale piece of sample from the site-specific area (e.g. at an interface) and then thin it for cross-sectional TEM observation. A FIB/TEM compatible in-situ heating holder was thus developed to emphasize this advantage. Figure 3.27 shows pictures of such a holder used by Tanigagaki et al. [171]. At the holder tip, a tungsten heating filament is used for the in-situ heating purpose. The holder was first loaded into a FIB system for preparation of a microscale site-specific cross-sectional sample containing an interface between a Mo film and Ta substrate. A lift-out piece was mounted onto a heating filament and the knob at the holder handle was turned to the "FIB" position in order to thin the sample by a 40 kV Ga ion beam (Fig. 3.27a). Once the sample was thinned and ready for TEM observation, the holder was pulled out from the FIB system, the knob at the holder handle was turned to the "TEM" position and then the holder was loaded into a transmission electron microscope to do in-situ heating TEM study on the Mo/Ta interface [171].

Improving resolution has always been a major pursue on the roadmap of electron microscopy and instrument developments, and in-situ TEM has been following the same trend. Along with the efforts to bring various external fields or forces to the TEM specimen area, in-situ TEM at atomic resolution is never forgotten, although such a high resolution is not always feasible. Some key requirements must be met by both electron microscopes and specimen holders in order to achieve atomic resolution in-situ TEM.

FIB position

(a)

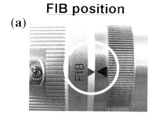

TEM position

(b)

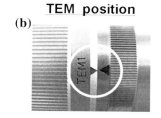

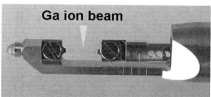

Ga ion beam

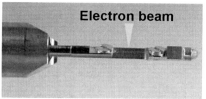

Electron beam

Fig. 3.27 In-situ FIB/TEM specimen holder allows preparation of a microscale site-specific cross-sectional TEM sample in a FIB system with the controller turned to position (**a**), the FIB position. After sample preparation by FIB, the controller of the holder is turned to position (**b**), the TEM position, and the same holder is then used to perform in-situ heating TEM (Patent of Hitachi High Technologies Corporation, Japan)

For the electron microscope, a large pole piece gap provides spacious room for inserting devices, illumination or field sources, or manipulators. However, the TEM image resolution will be compromised. For ETEM with gas injection ports on the pole piece, the electro-magnetic field within the pole piece may become asymmetric which lowers the image resolution, too. An optimized pole piece gap is crucial if atomic resolution in-situ TEM is a target. Additional pumps are often necessary for ETEM systems, therefore suitable anti-vibration system should be considered to reduce the vibration. Gas injection rate, total gas pressure or liquid amount surrounding a specimen should also be well controlled to minimize the influence on resolution.

For the in-situ TEM specimen holders, rule number 1 is simplicity. The design, arrangement of devices or field sources, sample fixation, wires, and pipelines should all be made or arranged in a way as simple as possible. For heating holders, it is extremely important to minimize or if possible eliminate the support grid, which usually contributes a major part of sample drift at temperature change. Keeping the power supply low and field- or force-application area local are proven to be efficient for lowering the sample drift.

3.4.2 Effects of Electron Beam Irradiation

As charged particles, electrons interact with matter much stronger than neutral X-rays or neutrons do. One of the often asked questions or discussed topics is the electron beam irradiation effects on phenomena observed during in-situ TEM experiments. The question is natural because in-situ TEM experiments are different from routine TEM observation. in-situ TEM often concentrates on one specific area and the observation may easily take a few tens of minutes or hours. Also, in order to continuously record the structural evolution, electron beam intensity must be kept high enough to maintain suitable imaging conditions. Such a long time electron beam irradiation itself may introduce more considerable changes in the materials compared with normal TEM experiments. How to decouple the beam effects from applied field effects is always a challenge.

Electron beam irradiation effects are complex. Heating (energy transfer), ionization (radiolysis, breaking bond), knock-on or displacement (knock atoms out of their lattice sites), electrostatic charging (electron transfer), and sputtering (remove atoms from surfaces) are main damage mechanisms [172]. Ionization damage may cause severe problems for organic or oxide materials but less problematic for metallic or semiconductor materials frequently studied with in-situ TEM because the ionized electrons can be rapidly replaced. Electron beam heating effect often raises a question about the real temperature at the specimen, although the temperature in the specimen chamber can be derived or directly measured [48]. Depending on materials, beam current, thermal dissipation, and other factors, the rise in temperature in a TEM specimen may range from a few to more than a hundred degrees Celsius which sometimes is high enough to cause visible structural changes such as motion of dislocations and defects [173–176], or phase transformations at much lower than nominal temperatures [177]. For the liquid TEM, the interaction between the high energy electron beam and the chemical solution produces species and/or hydrated electrons which may play an important role in the observed nanocrystal growth or interaction behaviors.

To know how to reduce the electron beam damage to certain materials is nontrivial, because the damage mechanisms are usually interrelated and material dependent. One popular way is to adjust the accelerating voltage of the electron beam. Increasing voltage reduces the thermal and ionization damages but enhances the knock-on damage and the surface sputtering and vise versa. Depending on materials to study, one must choose an optimum accelerating voltage to minimize the combined beam damages. For example, using 40–60 kV accelerating voltage for imaging carbon materials (the knock-on displacement threshold for carbon is between 27 and 95 kV [178]), 74 kV or lower for BN nanotubes [179], and 200 kV or lower for Si can largely minimize the knock-on damage. In combination with the favorable beam energy, reducing as much as possible the beam current and shortening the beam illumination time should also be considered. Based on this consideration, electron microscope alignment and routine astigmatism correction should be done in an unimportant specimen area, and electron beam should be blocked when illumination is not absolutely necessary. Figure 3.28a shows a case of the knock-on damage by a 300 kV electron beam (15 A/cm^2) to an amorphous carbon shell surrounding a TiO_2 nanoparticle. The damage became significant after 10 min beam irradiation, therefore it was hard to study the graphitization of the amorphous carbon shell by in-situ heating TEM. To shorten the exposure time, the electron beam was blocked after taking an initial

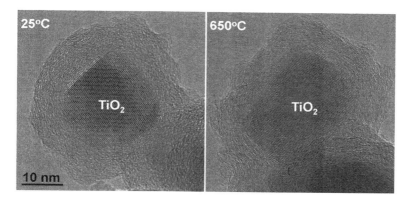

Fig. 3.28 A practical example of reducing electron beam irradiation effects in an in-situ TEM experiment. **a** As-prepared TiO_2 nanoparticle wrapped with an amorphous carbon shell. The image was taken at room temperature in a 300 kV H-9500 transmission electron microscope. **b** The specimen was heated to 650 °C and held for 20 min with the electron beam blocked, and then beam exposure was resumed for imaging. Total electron beam irradiation time was no longer than 5 min (the particles shown in two images are not the same ones.)

image at room temperature. The temperature was then raised to 600 °C using an in-situ heating TEM holder followed by resuming the electron beam illumination to image the structural change. It was clear that the carbon shell was graphitized by the thermal field (Fig. 3.28b) rather than electron beam irradiation.

Figure 3.7 shows a similar example, in which a SiC sample was heated for 25 h in a transmission electron microscope. Because the electron beam was shut off, the observed crystallization of the amorphous intergranular film was clearly independent of the electron beam effect.

Good conductive contacts between specimen and support or TEM holder also enhance thermal dissipation and reduce charging therefore minimizing the electron beam-induced temperature rise on the specimen and the charging effect. Cooling and beam shower (short time pre-illumination with a strong electron beam flux) help to retard or reduce the electron beam-induced sample damage and/or contamination process. Coating a layer of carbon on the sample surface may prevent sputtering damage. It was also reported that heating some specimens could heal the beam damages. Figure 3.29 shows a multi-walled carbon nanotube damaged by 200 kV electron beam (8 A/cm^2) irradiation for 20 min at room temperature. The corresponding EELS spectrum shows a broadened graphite σ* peak. Heating the sample to 600 °C largely decreased such beam damage as evidenced in the TEM image and the well-defined σ* peak in the corresponding EELS spectrum. It was actually a self-healing process in the damaged carbon regions.

Similar healing phenomena under irradiation of a 200 kV electron beam were found to cause superplasticity in amorphous SiO_2 nanowires and nanoparticles [119] and to improve the fracture strength of carbon nanotubes [112]. Heating may also cut down the hydrocarbon contamination rate by a factor of 10–30 depending on temperatures used [175].

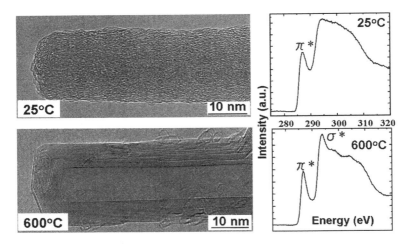

Fig. 3.29 Irradiation effects of a 200 kV electron beam on a multi-walled carbon nanotube. At room temperature, the electron beam deformed the graphite-like structure on the nanotube walls. EELS spectrum shows a broadened carbon σ^* peak. Under the same electron irradiation condition, no distinguishable structural deformation on the nanotube walls was recognized at 600 °C, and EELS spectrum shows well-defined σ^* peak (Hitachi High Technologies Corporation, Japan)

No matter what beam energy, electron imaging conditions, and other counter-measures are chosen, electron beam irradiation-induced damages to specimens can never go to zero especially after hours of in-situ TEM observations. In fact, it is often seen that the structural changes stem from a combined effect of the applied fields and the electron beam irradiation [119]. Caution is always needed in data interpretation. It should be a good habit to at least check the less or not irradiated neighboring specimen areas to make a comparison with the areas under an intensive and long-time electron beam illumination in order to understand and exclude the electron beam effects.

It is interesting to note that in some reported in-situ TEM experiments, electron beam irradiation effects were intentionally used to initiate dynamic processes. For example, electron beam irradiation was utilized to create knock-on displacement of carbon in multi-walled carbon nanotubes at 600 °C and the shrinkage of the carbon nanotubes resulted in a high pressure on the Fe_3C crystal(s) encapsulated inside the carbon nanotubes [124]. Similarly, carbon onions showed self-compression, too, under electron beam irradiation at ~700 °C [180]. Another experiment was about the role of electron beam irradiation on electric field evaporation of carbon nanotubes. The onset field for evaporation to start was lowered by about 30 % when the electron beam (200 kV, 100 A/cm^2 beam current density) was turned on, and the knock-on effect was believed to be the main reason [89]. Going back to Sect. 3.3.1, a 300 kV electron beam with a beam current density of 2–14 A/cm^2 transferred electrons to the Pt(acetylacetonate)$_2$ liquid solution and reduced Pt^{2+} to Pt [66]. Some beam effects are more significant with high gas pressure in gas ETEM. For example, gas molecules may be ionized by the electron beam, making the molecules more reactive to interact or etch solid materials. The electron beam may also have a potential to be used as electron tweezers to manipulate defects in materials.

3.4.3 Digital Recording

A picture is worth a thousand words, a movie tells a story. The in-situ TEM is about discovering dynamic responses of materials to the externally applied stimuli. In most cases the responses include the structural evolution and many critical structural changes such as phase transformation or melting often take place suddenly without a clear clue beforehand. Therefore, knowing how, and having a live TEM observation capability is only the first step, recording what is seen at an adequate time resolution is equally important. The conventional film camera takes at least 10 s (including film transportation) to complete one exposure, this speed is orders of magnitude slower than that of TV cameras (30 frames/s), and certainly cannot satisfy the needs for instantaneous exposure and continuous high-speed image recording. Video recording for the entire in-situ TEM process is the best way to guarantee the capture of any critical moment and the continuous evolution processes in materials.

Before digital CCD (charge-coupled device) cameras became prevalent in TEM applications in the 1990s, analog TV systems were used starting in the 1980s. Electron microscopes output TV signals in standard video formats (NTSC: 525 lines/50 Hz, or PAL: 625 lines/60 Hz) which allowed users to view the live images on a TV screen and record videos using standard video cassette recorders (VCR). The standard frame rate was 29.97 frames/s, or 33 ms per frame. The large information capacity of the magnetic cassette tapes could support up to 10 h of video recording, reviewing the recorded movies was pretty easy. The problems were editing and analysis. The video file recorded on a magnetic tape needed to be converted to digital using a special and costly instrument in order to clip videos, grab single still frames and add texts or other necessary illustrations. Also, the recorded video frames were not quantitative in terms of image intensity linearity. Other problems for the TV cameras included uneven background noise, poor dynamic range (only ~10^2), a few tens of times smaller field of view than that of the TEM film cameras, and non-square pixels [181].

The 21st century is the digital century with no exception for cameras used for TEM. Digital CCD cameras are currently used on all state-of-the-art transmission electron microscopes, darkrooms for film development and image printing have become history in many laboratories. The high sensitivity, high dynamic range, high image intensity linearity, correctable background noise, and convenience for frame analysis and editing make the CCD cameras ideal for in-situ TEM. Via software, continuous digital image output from electron microscopes can be captured and made into movie files with the formats of avi, wmv, mpeg, etc. It should be noted that the video referred here is defined as a continuous stream of images in time, therefore not necessarily to be NTSC or PAL format. A wide range of CCD recording frame rates are commercially available, and 10–30 frames/s are typically desired for the purpose of in-situ TEM. Higher frame rates are certainly desirable but the shortened exposure time for each frame lowers the image quality. The frame size in digital video recording varies but is typically in the range of 1024 pixel × 1024 pixel or smaller as of today yielding to the high speed.

The recent progress in developing CMOS (complementary metal oxide semiconductor) imaging sensors sheds a light on high speed and high sensitivity digital streaming video recording. Because no charge transfer will be necessary, each pixel on the detector chip takes and processes electrons simultaneously on their own, this is different from the working principle of the CCD cameras. Digital TEM cameras with a frame rate of 1000 frames/s will become a reality in the near future although there is still a concern about whether or not the number of electrons recorded on each frame is adequate to present images well. Another concern is that if computer processing and image saving speeds cannot keep pace with the frame capture rate, frames will be dropped from time to time resulting in sacrificed time resolution and information lost. With the advent of digital video recording technology, high computer processing speed and associated information technologies, such a frame drop problem is possible to be avoided. Large hard disk capacity is a must for the live digital video recording which easily generates 20 MB/s or more information to be saved, 500–1000 GB capacity should be available for each in-situ TEM experiment with a typical 2–4 h total time of video recording. Sometimes multiple hard disks are used to save image frames so that frame dropping will not happen.

3.4.4 Technical Challenges Ahead

In the past more than half century, in-situ TEM technology development mainly focused on developing capabilities of applying fields and/or environments to the specimen area while performing TEM observation. The attention now moves to pursue the atomic spatial resolution and high temporal resolution. High priority will also be given to quantitative in-situ TEM technologies in the near future. Although no one will argue that in-situ TEM is entering into a booming time, continuous and innovative efforts are required to conquer technical and engineering challenges that lie ahead, these include but are not limited to:

– How to precisely control and quantitatively measure the applied fields or exotic environments at TEM specimens, for example temperature, gas pressure, gas or liquid composition right on TEM specimens?
– How to enable the atomic resolution TEM imaging in versatile environments applied to TEM specimen chambers?

- How to achieve sub-nanometer or higher imaging resolution for the in-situ liquid E-cell TEM? Chromatic aberration-corrected TEM is certainly one of the solutions although the cost is high.
- How to squeeze multiple external fields or illumination sources into the narrow pole piece gap of a transmission electron microscope while maintain high resolution power? A large pole piece gap combined with aberration-corrected TEM is promising.
- How to observe, capture, and store high quality images at high temporal resolution on a level of microseconds to picoseconds? The pulsed laser beam-induced ultrafast TEM (see Chap. 4) may make a breakthrough in this aspect.
- How to analyze the recorded amount of digital data in a quantitative and efficient time?
- How to identify the side effects of electron beam irradiation and to reduce their influences?
- How to deal with beam-sensitive materials? This is a particularly tough challenge for in-situ TEM from the technical point of view.

3.5 Summary

Depending on what external fields or forces that are applied to the specimen in a transmission electron microscope, one can have in-situ heating TEM, in-situ cooling TEM (LN_2 and LHe), in-situ environmental TEM (gas or liquid), in-situ biasing TEM (electric voltage and current), in-situ straining TEM (tension or compression), in-situ Lorentz TEM, in-situ electron holography TEM, and in-situ ion beam (or laser) irradiation TEM available for study of materials in stimulating environments. Real-time in-situ TEM observations at high spatial and time resolutions for dynamic structural processes are becoming routine nowadays. In most cases, the in-situ TEM capability can be realized on any standard transmission electron microscopes, and all that needs to be done is to design a special TEM specimen holder, which supplies a desired stimulus to the TEM specimen. These are holder-based in-situ TEM technologies. Column-based technologies like in-situ ETEM and in-situ ion beam irradiation TEM are more favorable when high resolution and multiple external stimuli are desired. The reported in-situ TEM experimental results have shown great impact on materials science and nanoscience and nanotechnology. New insights have been obtained for fundamental microscopic processes, structure-property relationships, catalytic mechanisms, and mechanisms governing materials properties. The more than ever important role of the in-situ TEM in making breakthroughs in scientific and technological R&Ds is stimulating another wave of advancement in electron microscopy after the aberration-corrected TEM.

Acknowledgments Section 3.5.4 was written with major contributions from X. Han, Y. Zhang, K. Zheng, Y. Yue, and Z. Zhang. The author thanks Emily Zhang for helping proof reading.

References

1. NSF Workshop Report on Dynamic in situ electron microscopy as a tool to meet the challenges of the nanoworld, January 3–6, 2006, The Buttes, Tempe, Arizona
2. B.J. Ford, *The Leeuwenhoek Legacy* (Biopress, Bristol, 1991)
3. E. Ruska, Beitrag zur übermikroskopischen Abbildung bei höheren Drucken. Kolloid Z. **100**, 212 (1942)
4. J. Silcox, M.J. Whelan, Direct observations of the annealing of prismatic dislocation loops and of climb of dislocations in quenched aluminum. Philos. Mag. **5**, 1 (1960)
5. N.H. Packan, D.N. Braski, Electron microscope in situ annealing study of voids induced by irradiation in aluminum. J. Nucl. Mater. **34**, 307 (1970)
6. R.T.K. Baker, In situ electron microscopy studies of catalyst particle behavior. Catal. Rev. **19**, 161 (1979)
7. ASU electron microscopy workshop. Microsc. Microanal. **4** (1998)
8. X.F. Zhang, L.C. De Jonghe, Thermal modification of microstructures and grain boundaries in silicon carbide. J. Mater. Res. **18**(12), 2807 (2003)
9. M. Law, X.F. Zhang, R. Yu, T. KuyKendall, P. Yang, Thermally driven interfacial dynamics of metal/oxide bilayer nanoribbons. Small **1**, 1 (2005)
10. R. Yu, H. Song, X.F. Zhang, P. Yang, Thermal wetting of platinum nanocrystals on silica surface. J. Phys. Chem. B **109**, 6940 (2005)
11. S.H. Oh, Y. Kauffmann, C. Scheu, W.D. Kaplan, M. Ruhle, Ordered liquid aluminum at the interface with sapphire. Science **310**, 661 (2005)
12. S.K. Eswaramoorthy, J.M. Howe, G. Muralidharan, In situ determination of the nanoscale chemistry and behavior of solid-liquid systems. Science **318**, 1437 (2007)
13. V.C. Holmberg, M.G. Panthani, B.A. Korgel, Phase transitions, melting dynamics, and solid-state diffusion in a nano test tube. Science **326**, 405 (2009)
14. T. Kamino, H. Saka, A newly developed high resolution hot stage and its application to materials characterization. Microsc. Microanal. Microstruct. **4**, 127 (1993)
15. T. Kamino, T. Yaguchi, H. Saka, In situ study of chemical reaction between silicon and graphite at 1,400°C in a high resolution/analytical electron microscope. J. Electron Microsc. (Tokyo), **43**, 104 (1994)
16. T. Kamino, H. Saka, In-situ HREM heating experiments at very high temperatures. Mat. Res. Soc. Symp. Proc. **404**, 19 (1996)
17. S. Tsukimoto, S. Arai, M. Konno, T. Kamino, K. Sasaki, H. Saka, In situ high resolution electron microscopy/electron energy loss spectroscopy observation of wetting of a Si surface by molten Al. J. Microsc. **203**, 17 (2001)
18. J.G. Lee, H. Mori, In-situ observation of alloy phase formation in nanometre-sized particles in the Sn-Bi system. Philos. Mag. **84**, 2675 (2004)
19. T. Kamino, T. Yaguchi, M. Konno, T. Hashimoto, In situ high temperature TEM observation of interaction between multi-walled carbon nanotube and in situ deposited gold nano-particles. J. Electron Microsc. **54**, 461 (2005)
20. T. Kamino, T. Yaguchi, T. Sato, H. Hashimoto, Development of a technique for high resolution electron microscopic observation of nano-materials at elevated temperatures. J. Electron Microsc. **54**, 505 (2005)
21. T. Akita, K. Tanaka, M. Kohyama, M. Haruta, Analytical TEM study on structural changes of Au particles on cerium oxide using a heating holder. Catal. Today **122**, 233 (2007)
22. H. Saka, T. Kamino, S. Arai, K. Sasaki, In situ heating transmission electron microscopy. MRS Bull. **33**, 93 (2008)
23. T. Tanigagaki, K. Ito, Y. Nagakubo, T. Asakawa, T. Kanemura, An in situ heating TEM analysis method for an interface reaction. J. Electron Microsc. **58**, 281 (2009)
24. T. Kamino, T. Yaguchi, M. Lonno, A. Watabe, Y. Nagakubo, Development of a specimen heating holder with an evaporator and gas injector and its application for catalyst. J. Electron Microsc. **55**, 245 (2006)

25. L.F. Allard, W.C. Biglow, M. Jose-Yacaman, D.P. Nackashi, J. Damiano, S. Mick, A new MEMS-based system for ultra-high-resolution imaging at elevated temperatures. Microsc. Res. Technol. **72**, 208 (2009)

26. Private communication with *Fischione* Instruments

27. T. Kamino, T. Yaguchi, M. Konno, A. Watabe, T. Marukawa, T. Mima, K. Kuroda, H. Saka, S. Arai, H. Makino, Y. Suzuki, K. Kishita, Development of a gas injection/specimen heating holder for use with transmission electron microscope. J. Electron Microsc. **54**, 497 (2005)

28. A. Tonomura, Direct observation of thitherto unobservable quantum phenomena by using electrons. PNAS **102**, 14952 (2005)

29. J. Frank, *Electron Tomography—Three Dimensional Imaging with the Transmission Electron Microscope* (Plenum Press, New York, 1992)

30. K. Dierksen, D. Typke, R. Hegerl, A.J. Koster, W. Baumeister, Towards automatic electron tomography. Ultramicroscopy **40**, 71 (1992)

31. K.H. Downning, H. Sui, M. Auer, Electron tomography: a 3D view of the subcellular world. Anal. Chem. **11**, 7949 (2007)

32. A.M. Glauert, The high voltage electron microscope in biology. J. Cell Biol. **63**, 717 (1974)

33. L. Marton, Bull. Acad. R. Belg. C1. Sci. **21**, 553 (1935)

34. E.P. Butler, K.F. Hale, *Dynamic Experiments in the Electron Microscope* (North-Holland, Amsterdam, 1981)

35. T. Yaguchi, A. Watanabe, Y. Nagakubo, K. Ueda, M. Fukui, T. Kamino, T. Kawasaki, Development of gas environmental cells for in situ TEM applications, in *Proceedings of Microscopy and Microanalysis Annual Meeting*, 8/1-5, 2010, Portland

36. M.J. Flower, High voltage electron microscopy of environmental reactions. J. Microsc. **97**, 171 (1973)

37. P.R. Swann, N.J. Tighe, Jernkont. Annlr. **155**, 251 (1971)

38. P.R. Swann, N.J. Tighe, Performance of differentially pumped environmental cell in the AE1 EM7, in *Proceeding of 5th European Congress on Electron Microscopy*, Manchester (1972), p. 360

39. P.R. Swann, High voltage microscope studies of environmental reaction, in *Electron Microscopy and Structure of Materials*, ed. by G. Thomas, R. Fulrath, R.M. Fisher (University of California Press, Berkeley, 1972), p. 878

40. T.C. Lee, D.K. Dewald, J.A. Eades, I.M. Robertson, H.K. Birnbaum, An environmental cell transmission electron microscope. Rev. Sci. Instrum. **62**, 1438 (1991)

41. E.D. Boyes, P.L. Gai, Environmental high resolution electron microscopy and applications to chemical science. Ultramicroscopy **67**, 219 (1997)

42. R. Sharma, K. Weiss, Development of a TEM to study in situ structural and chemical changes at an atomic level during gas-solid interactions at elevated temperatures. Microsc. Res. Techniq. **42**, 270 (1998)

43. P.L. Gai, E.D. Boyes, S. Helveg, P.L. Hansen, S. Giorgio, C.R. Henry, Atomic-resolution environmental transmission electron microscopy for probing gas-solid reactions in heterogeneous catalysis. MRS Bull. **32**, 1044 (2007)

44. K. Kishita, H. Sakai, H. Tanaka, H. Saka, K. Kuroda, M. Sakamoto, A. Watabe, T. Kamino, Development of an analytical environmental TEM system and its application. J. Electron. Microsc. **58**, 331 (2009)

45. P.L. Gai, R. Sharma, F.M. Ross, Environmental (S)TEM studies of gas-liquid-solid interactions under reaction conditions. MRS Bull. **33**, 107 (2008)

46. X.F. Zhang, T. Kamino, Imaging gas-solid interactions in an atomic resolution environmental TEM. Microsc. Today **14**, 16 (2006)

47. X.F. Zhang, In-situ gas-heating and real-time 3D imaging: Hitachi H-9500 transmission electron microscope. Am. Lab. **40**, 27 (2008)

48. P.A. Crozier, R. Sharma, A.K. Datye, Oxidation and reduction of small palladium particles on silica. Microsc. Microanal. **4**, 278 (1998)

49. T.W. Hansen, J.B. Wagner, P.L. Hansen, S. Dahl, H. Topsee, C.J.H. Jacobsen, Atomic-resolution in situ transmission electron microscopy of a promoter of a heterogeneous catalyst. Science **294**, 1508 (2001)

50. S. Helveg, C. Lopez-Cartes, J. Hehested, P.L. Hansen, B.S. Calusen, J.R. Rostrup-Nielsen, F. Abild-Pedersen, J.K. Nerskov, Atomic-scale imaging of carbon nanofibre growth. Nature **427**, 426 (2004)
51. J.B. Hannon, S. Kodambaka, F.M. Ross, R.M. Tromp, The influence of the surface migration of gold on the growth of silicon nanowires. Nature **440**, 69 (2006)
52. S. Kodambaka, J. Tersoff, M.C. Reuter, F.M. Ross, Germanium nanowire growth below the eucentric temperature. Science **316**, 729 (2007)
53. B.J. Kim, J. Tersoff, S. Kodambaka, M.C. Reuter, E.A. Stach, F.M. Ross, Kinetics of individual nucleation events observed in nanoscale vapor-liquid-solid growth. Science **322**, 1070 (2008)
54. S. Hofmann, R. Sharma, C.T. Wirth, F. Cervantes-Sodi, C. Ducati, T. Kasama, R.E. Dunin-Borkowski, J. Drucker, P. Bennett, J. Robertson, Ledge-flow-controlled catalyst interface dynamics during Si nanowire growth. Nat. Mater. **7**, 372 (2008)
55. V.P. Oleshko, P.A. Crozier, R.D. Cantrell, A.D. Westwood, J. Electron Microsc. **51**(supplement), S27 (2002)
56. R. Sharma, P. Rez, M. Brown, G.H. Du, M.M.J. Treacy, Nanotechnology **18**, 125602 (2007)
57. M. Haider, H. Rose, S. Uhlemann, B. Kabius, K. Urban, Nature **392**, 768 (1998)
58. N. Dellby, O.L. Krivanek, P.D. Nellist, P.E. Batson, A.R. Lupini, J. Electron Microsc. (Tokyo) **50**, 177 (2001)
59. I.M. Abrams, J.W. McBain, Science **100,** 273 (1944)
60. I.M. Abrams, J.W. McBain, J. Appl. Phys. **15,** 607 (1944)
61. R.T. Joy, The electron microscopical observation of aqueous biological systems. Adv. Opt. Electron Microsc. **5**, 297 (1973)
62. D.F. Parsons, Environmental wet cells for biological medium voltage and high voltage electron microscopy, in *Electron Microscopy* (1973)
63. B. Siegel (ed.), in *Physical Aspects* (Wiley, New York, 1974)
64. D.F. Parsons, Structure of wet specimens in electron microscopy. Science **186**, 407 (1974)
65. M.J. Williamson, R.M. Tromp, P.M. Vereecken, R. Hull, F.M. Ross, Dynamic microscopy of nanoscale cluster growth at the solid-liquid interface. Nat. Mater. **2**, 532 (2003)
66. H. Zheng, R.K. Smith, Y. Jun, C. Kisielowski, U. Dahmen, A.P. Alivisatos, Observation of single colloidal platinum nanocrystal growth trajectories. Science **324**, 1309 (2009)
67. H. Liao, L. Cui, S. Whitelam, H. Zheng, Real-time imaging of Pt_3Fe nanorod growth in solution. Science **336**, 1011 (2012)
68. J.D. Li, M.H. Nielsen, J.R. Lee, C. Frandsen, J.F. Banfield, J. De Yoreo, Direction-specific interactions control crystal growth by oriented attachment. Science **336**, 1014 (2012)
69. J.M. Yuk, J. Park, P. Ercius, K. Kim, D.J. Hellebusch, M.F. Crommie, J.Y. Lee, A. Zettl, A. Alivisatos, High-resolution EM of colloidal nanocrystal growth using graphene liquid cells. Science **336**, 61 (2012)
70. N. de Jonge, D.B. Peckys, G.J. Kremers, D.W. Piston, Electron microscopy of whole cells in liquid with nanometer resolution. Proc. Natl. Acad. Sci. USA **106**(7), 2159–2164 (2009). Epub 2009 Jan 21
71. D.B. Peckys, G.M. Veith, D.C. Joy, N. de Jonge, Nanoscale imaging of whole cells using a liquid enclosure and a scanning transmission electron microscope. Plos One **4**, e8214 (2009)
72. R.C. Moretz, G.G. Hausner, JR., D.F. Parsons, in *Proceedings of the 29th Annual Meeting of the Electron Microscopy Society of America,* Boston, ed. by C.J. Arceneaux (Claitor's, Baton Rouge, 1971), p. 544
73. S.W. Hui, D.F. Parsons, in *Proceedings of the 31th Annual Meeting of the Electron Microscopy Society of America,* New Orleans, ed. by C.J. Arceneaux (Claitor's, Baton Rouge, 1973), p. 340
74. P.L. Gai, Development of wet environmental TEM (Wet-ETEM) for in situ studies of liquid-catalyst reactions on the nanoscale. Microsc. Microanal. **8**, 21 (2002)
75. P.W. Sutter, E.A. Sutter, Dispensing and surface-induced crystallization of aeptolitre liquid metal-alloy drops. Nat. Mater. **6**, 363 (2007)
76. J.Y. Huang, S. Ceh, Z.Q. Wang, K. Kempa, Y.M. Wang, S.H. Jo, G. Chen, M.S. Dresselhaus, Z.F. Ren, Superplastic carbon nanotubes. Nature **439**, 281 (2006)

77. J.Y. Huang, S. Chen, Z.F. Ren, Z. Wang, K. Kempa, M.J. Naughton, G. Chen, M.S. Dresselhaus, Enhanced ductile behavior of tensile-elongated individual double-walled and triple-walled carbon nanotubes at high temperatures. Phys. Rev. Lett. **98**, 185501-1 (2007)
78. J.Y. Huang, S. Chen, Z.F. Ren, G. Chen, M.S. Dresselhaus, Real-time observation of tubule formation from amorphous carbon nanowires under high-bias Joule heating. Nano Lett. **6**, 1699 (2006)
79. B.C. Regan, S. Aloni, R.O. Ritchie, U. Dahmen, A. Zettl, Carbon nanotubes as nanoscale mass conveyors. Nature **428**, 924 (2004)
80. A. Barreiro, R. Rurali, E.R. Hernandez, J. Moser, T. Pichler, L. Forro, A. Bachtold, Subnanometer motion of cargoes driven by thermal gradiants along carbon nanotubes. Science **320**, 775 (2008)
81. J. Cumings, P.G. Collins, A. Zettl, Peeling and sharpening multiwall nanotubes. Nature **406**, 586 (2000)
82. J. Cumings, A. Zettl, Low-friction nanoscale linear bearing realized from multiwall carbon nanotubes. Science **289**, 602 (2000)
83. P. Gao, Z. Kang, W. Fu, W. Wang, X. Bai, E. Wang, Electrically driven redox process in cerium oxides. J. Am. Chem. Soc. **132**, 4197 (2010)
84. P. Poncharal, C. Berger, Y. Yi, Z.L. Wang, W.A. de Heer, Room temperature ballistic conduction in carbon nanotubes. J. Phys. Chem. B **106**, 12104 (2002)
85. M.S. Wang, J. Wang, Q. Chen, L.M. Peng, Fabrication and electrical and mechanical properties of carbon nanotube interconnections. Adv. Funct. Mater. **15**, 1825 (2005)
86. M.S. Wang, L.M. Peng, J.Y. Wang, Q. Chen, Shaping carbon nanotubes and effects on their electrical and mechanical properties. Adv. Funct. Mater. **16**, 1462 (2006)
87. K.H. Liu, P. Gao, Z. Xu, X.D. Bai, E.G. Wang, In situ probing electrical response on bending of ZnO nanowires inside transmission electron microscope. Appl. Phys. Lett. **92**, 213105-1 (2008)
88. M.S. Wang, Q. Chen, L.M. Peng, Field-emission characteristics of individual carbon nanotubes with a conical tip: the validity of the Fowler-Nordheim theory and maximum emission current. Small **4**, 1907 (2008)
89. M.S. Wang, Q. Chen, L.M. Peng, Grinding a nanotube. Adv. Mater. **20**, 724 (2008)
90. A.N. Chiaramonti, L.J. Thompson, W.F. Egelhoff, B.C. Kabius, A.K. Petford-Long, In situ TEM studies of local transport and structure in nanoscale multilayer films. Ultramicroscopy **108**, 1529 (2008)
91. A.N. Chiaramonti, D.K. Schreiber, W.F. Egelhoff, D.N. Seidman, A.K. Petford-Long, Effects of annealing on local composition and electrical transport correlations in MgO-based magnetic tunnel junctions. Appl. Phys. Lett. **93**, 103113-1 (2008)
92. J.W. Lau, P. Morrow, J.C. Read, V. Höink, W.F. Egelhoff, L. Huang, Y. Zhu, In situ tunneling measurements in a transmission electron microscope on nanomagnetic tunnel junctions. Appl. Phys. Lett. **96**, 262508-1 (2010)
93. J.Y. Huang, L. Zhong, C.M. Wang, J.P. Sullivan, W. Xu, L.Q. Zhang, S.X. Mao, N.S. Hudak, X.H. Liu, A. Subramanian, H. Fan, L. Qi, J. Li, In situ observation of the electrochemical lithiation of a single SnO_2 nanowire electrode. Science **330**, 1515 (2010)
94. T. Fujieda, K. Hidaka, M. Hayashibara, T. Kamino, H. Matsumoto, Y. Ose, H. Abe, T. Shimizu, H. Tokumoto, In situ observation of field emission from an individual carbon nanotube by Lorenz microscopy. Appl. Phys. Lett. **85**, 5739 (2004)
95. T. Fujieda, K. Hidaka, M. Hayashibara, T. Kamino, Y. Ose, H. Abe, T. Shimizu, H. Tokumoto, Direct observation of field emission sites in the single multiwalled carbon nanotube by Lorenz microscopy. Jpn. J. Appl. Phys. **44**, 1661 (2005)
96. T. Fujieda, M. Okai, K. Hidaka, H. Matsumoto, H. Tokumoto, Behavior of catalyst particle at tip of carbon nanotube during field emission. Appl. Phys. Express **1**, 014002-1 (2008)
97. K. Liu, W. Wang, Z. Xu, X. Bai, E. Wang, Y. Yao, J. Zhang, Z. Liu, Chirality-dependent transport properties of double-walled nanotubes measured in situ on their field-effect transistors. J. Am. Chem. Soc. **131**, 62 (2009)
98. Z.W. Shan, G. Adesso, A. Cabot, M.P. Sherburne, S.A. Syed Aasif, O.L. Warren, D.C. Chrzan, A.M. Minor, A.P. Alivisatos, Ultrahigh stress and strain in hierarchically structured hollow nanoparticles. Nat. Mater. **7**, 947 (2008)

99. N. Gane, F.P. Bowden, J. Appl. Phys. **39**, 1432 (1968)

100. N. Gane, Proc. R. Soc. Lond. Ser. A **317**, 367 (1970)

101. M. Legros, D.S. Gianola, C. Motz, Quantitative in situ mechanical testing in electron microscopes. MRS Bull. **35**, 354 (2010)

102. P. Poncharal, Z.L. Wang, D. Ugart, W.A. de Heer, Electrostatic deflections and electromechanical resonances of carbon nanotubes. Science **283**, 1513 (1999)

103. H.G.F. Wilsdorf, ASTM Spec. Technol. **245**, 43 (1958)

104. U. Messerschmidt, F. Appel, Ultramicroscopy **1**, 223 (1976)

105. E.P. Butler, Rep. Prog. Phys. **42**, 833 (1979)

106. I.M. Robertson, P.J. Ferreira, G. Dehm, R. Hull, E.A. Stach, Visualizing the behavior of dislocations—seeing is believing. MRS Bull. **33**, 122 (2008)

107. I.M. Robertson, H.K. Birnbaum, P. Sofronis, Hydrogen effects on plasticity, in *Dislocations in Solids*, ed. by J. P. Hirth, L. Kubin (Elsevier B.V, Netherlands, 2009), pp. 249–293

108. Z.W. Shan, E.A. Stach, J.M.K. Wiezorek, J.A. Knapp, D.M. Follstaedt, S.X. Mao, Grain boundary-mediated plasticity in nanocrystalline nickel. Science **305**, 654 (2004)

109. Z.W. Shan, J.M.K. Wiezorek, E.A. Stach, D.M. Follstaedt, J.A. Knapp, S.X. Mao, Dislocation dynamics in nanocrystalline nickel. Phys. Rev. Lett. **98**, 095502-1 (2007)

110. Z.W. Shan, J.A. Knapp, D.M. Follstaedt, E.A. Stach, J.M.K. Wiezorek, S.X. Mao, Inter- and intra-agglomerate fracture in nanocrystalline nickel. Phys. Rev. Lett. **100**, 105502-1 (2008)

111. M.A. Haque, H.D. Espinosa, H.J. Lee, MEMS for in situ testing—Handling, actuation, loading, and displacement measurements. MRS Bull. **35**, 375 (2010)

112. B. Peng, M. Locascio, P. Zapol, S.Y. Li, S.L. Mielke, G.C. Schatz, H.D. Espinosa, Measurements of near-ultimate strength for multiwalled carbon nanotubes and irradiation-induced crosslinking improvements. Nat. Nanotechnol. **3**, 626 (2008)

113. M. Locascio, B. Peng, P. Zapol, Y. Zhu, S.Y. Li, S.L. Mielke, T. Belytschko, H.D. Espinosa, Exp. Mech. **49**, 169 (2009)

114. Z.W. Shan, R.K. Mishra, S.A.S. Asif, O.L. Warren, A.M. Minor, Mechanical annealing and source-limited deformation in submicrometer-diameter Ni crystals. Nat. Mater. **7**, 115 (2007)

115. M.S. Wang, I. Kaplan-Ashirl, X.L. Wei, R. Rosentsveig, H.D. Wagner, R. Tenne, L.M. Peng, In situ TEM measurements of the mechanical properties and behavior of WS2 nanotubes. Nano Res. **1**, 22 (2008)

116. Q. Yu, Z.W. Shan, J. Li, X.X. Huang, L. Xiao, J. Sun, E. Ma, Strong crystal size effect on deformation twinning. Nature **463**, 335 (2010)

117. X.D. Han, K. Zheng, Y.F. Zhang, X.N. Zhang, Z. Zhang, Z.L. Wang, Low temperature in situ large strain plasticity of silicon nanowires. Adv. Mater. **19**, 2112 (2007)

118. Y.F. Zhang, X.D. Han, K. Zheng, Z. Zhang, X.N. Zhang, J.Y. Fu, Y. Ji, Y.J Hao, X.Y. Guo, Z.L. Wang, Direct observation of super-plasticity of beta-SiC nanowires at low temperature. Adv. Funct. Mater. **17**, 3435 (2007)

119. K. Zheng, C.C. Wang, Y.Q. Cheng, Y.H. Yue, X.D. Han, Z. Zhang, Z.W. Shan, S.X. Mao, M.M. Ye, Y.D. Yin, E. Ma, Electron-beam-assisted superplastic shaping of nanoscale amorphous silica. Nat. Commun. **1**, 1 (2010)

120. X.D. Han, Y.F. Zhang, K. Zheng, X.N. Zhang, Z. Zhang, Y.J. Hao, X.Y. Guo, J. Yuan, Z.L. Wang, Direct observation of super-plasticity of beta-SiC nanowires at low temperature. Nano Lett. **7**, 452 (2007)

121. K. Zheng, X.D. Han, L.H. Wang, Y.H. Yue, Y.F. Zhang, Y. Qin, X.N. Zhang, Z. Zhang, Atomic mechanisms governing the elastic limit and the incipient plasticity of bending Si nanowires. Nano Lett. **9**, 2471 (2009)

122. X.D. Han, Y.F. Zhang, X.Q. Liu, Z. Zhang, Y.J. Hao, X.Y. Guo, Lattice bending, disordering and amorphization induced plastic deformation in a SiC nanowire. J. Appl. Phys. **98**, 124307-1 (2005)

123. J.H. Wang, S. Yip, S.R. Phillpot, D. Wolf, Crystal instabilities at finite strain. Phys. Rev. Lett. **71**, 4182 (1993)

124. L. Sun, F. Banhart, A.V. Krasheninnikov, J.A. Rodriguez-Manzo, M. Terrones, P.M. Ajayan, Carbon nanotubes as high pressure cylinders and nanoextruders. Science **312**, 1199 (2006)

125. J. Cumings, E. Olsson, A.K. Petford-Long, Y. Zhu, Electric and magnetic phenomena studied by in situ transmission electron microscopy. MRS Bull. **33**, 101 (2008)

126. C. Tsuruta, T. Kamino, H. Sato, *Observation of magnetic domain in cobalt thin foil*, Hitachi Technical Data, Sheet No. 51

127. T. Hirayama, Q. Ru, T. Tanji, A. Tonomura, Observation of magnetic-domain states of barium ferrite particles by electron holography. Appl. Phys. Lett. **63**, 418 (1993)

128. J.J. Kim, A. Tonomura, K. Hirata, Y. Ishida, D. Shindo, M. Takahashi, Magnetic domain observation in writer pole tip for perpendicular recording head by electron holography. Appl. Phys. Lett. **92**, 162501 (2008)

129. A. Sugawara, K. Fukunaga, M.R. Scheinfein, H. Kobayashi, H. Kitagawa, A. Tonomura, Electron holography study of the temperature variation of the magnetic order parameter within circularly nickel nanoparticle rings. Appl. Phys. Lett. **91**, 262513 (2007)

130. A. Sugawara, T. Akashi, P.D. Brown, R.P. Campion, T. Yoshida, B.L. Gallagher, A. Tonomura, High-resolution observations of temperature-dependent magnetic domain structures within $Ga_xMn_{1-x}As$ by Lorentz microscopy. Phys. Rev. B **75**, 241306-1 (2007)

131. K. Fukunaga, A. Sugawara, Anisotropic cross-tie wall and their confinement in self organized undulating Fe film. J. Appl. Phys. **103**, 053909 (2008)

132. A. Sugawara, H. Kasai, A. Tonomura, P.D. Brown, R.P. Campion, K.W. Edmonds, B.L. Gallagher, J. Zemen, T. Jungwirth, Domain walls in the (Ga, Mn) As diluted magnetic semiconductor. Phys. Rev. Lett. **100**, 047202 (2008)

133. A.C. Twitchett, R.E. Dunin-Borkoski, P.A. Midgley, Philos. Mag. **86**, 5805 (2006)

134. A. Tonomura, *The Quantum World Unveiled by Electron Waves* (World Scientific, Singapore, 2008)

135. H. Lichte, M. Lehmann, Electron holography—basics and applications. Rep. Prog. Phys. **70**, 1 (2007)

136. R.E. Dunin-Boekowski, T. Kasama, A. Wei, S.L. Tripp, M.J. Hytch, E. Snoeck, R.J. Harrison, A. Putnis, Off-axis electron holography of magnetic nanowires and chains, rings, and plannar arrays of magnetic nanoparticles. Microsc. Res. Technol. **64**, 390 (2004)

137. Y. Togawa, K. Harada, T. Akashi, H. Kasai, T. Matsuda, F. Nori, A. Maeda, A. Tonomura, Direct observation of rectified motion of vortices in a niobium superconductor. Phys. Rev. Lett. **95**, 087002 (2005)

138. Y. Togawa, T. Kimura, K. Harada, T. Akashi, T. Matsuda, A. Tonomura, Y. Otani, Current-excited magnetization dynamics in narrow ferromagnetic wires. Jpn. J. Appl. Phys. (Express) **45**, L683 (2006)

139. Y. Togawa, T. Kimura, K. Harada, T. Akashi, T. Matsuda, A. Tonomura, Y. Otani, Current-excited magnetization reversal under in-plane magnetic field in a nano-scaled ferromagnetic wire. Appl. Phys. Lett. **92**, 012505 (2008)

140. Y. Togawa, K. Harada, T. Akashi, H. Kasai, T. Matsuda, A. Maeda, A. Tonomura, Rectified motion of vortices in a niobium superconductor observed by Lorentz microscopy. Physica C **426–431**, 141 (2005)

141. A. Tonomura, H. Kasai, O. Kamimura, T. Matsuda, K. Harada, T. Yoshida, T. Akashi, J. Shimoyama, K. Kishio, T. Hanaguri, K. Kitazawa, T. Masui, S. Tajima, N. Koshizuka, P.L. Gammel, D. Bishop, M. Sasase, S. Okayasu, Observation of structures of chain vortices inside anisotropic high-T_c superconductors. Phys. Rev. Lett. **88**, 237001 (2002)

142. R.E. Dunin-Borkowski, M.R. McCartney, D.J. Smith, S.S.P. Parkin, Towards quantitative electron holography of magnetic thin films using in situ magnetization reversal. Ultramicroscopy **74**, 61 (1998)

143. J.W. Lau, M. Beleggia, M.A. Schofield, G.F. Neumark, Y. Zhu, Direct correlation of reversal rate dynamics to domain configurations in micron-sized permalloy elements. J. Appl. Phys. **97**, 10E702-1 (2005)

144. J.W. Lau, M. Beleggia, Y. Zhu, Common reversal mechanisms and correlation between transient domain states and field sweep rate in patterned Permalloy structures. J. Appl. Phys. **102**, 043906-1 (2007)

145. Z. Akase, D. Shindo, In situ Lorentz microscopy in an alternating magnetic field. J. Electron Microsc. **59**, 207 (2010)

146. A. Lenk, H. Lichte, U. Muehle, J. Electron Microsc. **54**, 351 (2005)

147. S. Frabboni, G. Matteucci, G. Pozzi, M. Vanzi, Phys. Rev. Lett. **55**, 2196 (1985)

148. W.D. Rau, P. Schwander, F.H. Baumann, W. Hoppner, A. Ourmazd, Phys. Rev. Lett. **82**, 2614 (1999)

149. A.C. Twitchett, R.E. Dunin-Bokowski, P.A. Midgley, Phys. Rev. Lett. **88**, 238302 (2002)

150. A.C. Twitchett, R.E. Dunin-Bokowski, R.F. Broom, P.A. Midgley, J. Phys. Condens. Matter **16**, S181 (2004)

151. A.C. Twitchett-Harrison, T.J.V. Yates, S.B. Newcomb, R.E. Dunin-Borkowski, P.A. Midgley, Nano Lett. **7**, 2020 (2007)

152. A.C. Twitchett, R.E. Dunin-Borkowski, R.J. Hallifax, R.F. Broom, P.A. Midgley, Microsc. Microanal. **11**, 1 (2005)

153. X. Portier, E.Y. Tsymbal, A.K. Petford-Long, T.C. Anthony, J.A. Brug, Phys. Rev. B. **58**, R591 (1998)

154. F. Junginger, M. Klaui, D. Backes, U. Rudiger, T. Kasama, R.E. Dunin-Borkoski, L.J. Heyderman, C.A.F. Vaz, J.A.C. Bland, Appl. Phys. Lett. **90**, 132506 (2007)

155. M. Hayashi, L. Thomas, R. Moriya, C. Rettner, S.P. Parkin, Current-controlled magnetic domain-wall nanowire shift register. Science **320**, 209 (2008)

156. L. Huang, M.A. Schofield, Y. Zhu, Direct observation of the controlled magnetization reversal processes in Py/Al/Py asymmetric ring stacks. Appl. Phys. Lett. **95**, 042501-1 (2009)

157. L. Huang, M.A. Schofield, Y. Zhu, Control of double-vortex domain configurations in a shape-engineered trilayer nanomagnet system. Adv. Mater. **22**, 492 (2010)

158. T. Uhlig, M. Heumann, J. Zweck, Development of a specimen holder for in situ generation of pure in-plane magnetic fields in a transmission electron microscope. Ultramicroscopy **94**, 193 (2003)

159. M. Inoue, T. Tomita, M. Naruse, Z. Aakase, Y. Murakami, D. Shindo, Development of a magnetizing stage for in situ observations with electron holography and Lorentz microscopy. J. Electron Microsc. **54**, 509 (2005)

160. J.W. Lau, J.K. Bording, M. Beleggia, Y. Zhu, Energy barrier to magnetic vortex nucleation. Appl. Phys. Lett. **88**, 012508-1 (2006)

161. L. Huang, Y. Zhu, Controlled reversal of coupled Néel walls in flux-closure magnetic trilayer elements. Appl. Phys. Lett. **95**, 222502-1 (2009)

162. X.Z. Yu, Y. Onose, N. Kanazawa, J.H. Park, J.H. Han, Y. Matsui, N. Nagaosa, Y. Tokura, Real-space observation of a two-dimensional skyrmion crystal. Nature **465**, 901–904 (2010)

163. D. Shindo, Y.G. Park, Y. Gao, H.S. Park, Electron holography of Fe-based nanocrystalline magnetic materials. J. Appl. Phys. **95**, 6521 (2004)

164. H.S. Park, Y.G. Park, Y. Gao, D Shindo, M. Inoue, Direct observation of magnetization reversal in thin Nd2Fe14B film. J. Appl. Phys. **97**, 033908 (2005)

165. J.W. Lau, M.A. Schofield, Y. Zhu, A straightforward specimen holder modification for remnant magnetic-field measurement in TEM. Ultramicroscopy **107**, 396 (2007)

166. J. Cumings, A. Zettl, M.R. McCartney, J.C.H. Spence, Electron holography of field-emitting carbon nanotubes. Phys. Rev. Lett. **88**, 056804-1 (2002)

167. J. Cumings, A. Zettl, M.R. McCartney, Carbon nanotube electrostatic biprism: principle of operation and proof of concept. Microsc. Microanal. **10**, 420 (2004)

168. D.W. Pashley, A.E.B. Presland, Ion damage to metal films inside an electron microscope. Philos. Mag. **6**, 1003 (1961)

169. J.A. Hinks, A review of transmission electron microscopes with in situ ion irradiation. Nucl. Instrum. Methods Phys. Res. B **267**, 3652 (2009)

170. Y. Yan, Swift heavy ion irradiation damage in $YBa_2Cu_3O_{7-d}$ superconductors, in *Progress in Transmission Electron Microscopy, II. Applications in Materials Science*, ed. by X.F. Zhang, Z. Zhang (Springer/Tsinghua University Press, Berlin/Beijing, 2001), pp. 213–245

171. T. Tanigagaki, K. Ito, Y. Nagakubo, T. Asakawa, T. Kanemura, An in situ heating TEM analysis method for an interface reaction. J. Electron Microsc. **58**, 281 (2009)

172. R.F. Egerton, P. Li, M. Malac, Radiation damage in the TEM and SEM. Micron **35**, 399 (2004)

173. P.B. Hirsch, R.W. Horne, M.J. Whelan, Direct observations of the arrangement and motion of dislocations in aluminum. Philos. Mag. A **1**, 677 (1956)

174. P.B. Hirsch, Direct observations of moving dislocations: reflections on the thirtieth anniversary of the first recorded observations of moving dislocations by transmission electron microscopy. Mater. Sci. Eng. **84**, 1 (1986)

175. D.W. Pashley, M.J. Stowell, M.H. Jacobs, T.J. Law, The growth and structure of gold and silver deposits formed by evaporation inside an electron microscope. Philos. Mag. **10**, 127 (1964)

176. R. Sinclair, T. Yamashita, F.A. Ponce, Atomic motion on the surface of a cadmium telluride single crystal. Nature **290**, 386 (1981)

177. B.J. Kooi, JThM De Hosson, On the crystallization of thin films composed of $Sb_{3.6}Te$ with Ge for rewritable data storage. J. Appl. Phys. **95**, 4714 (2004)

178. O. Kamimura, T. Dobashi, K. Kawahara, T. Abe, K. Gohara, 10-kV diffractive Imaging using newly developed electron diffraction microscope. Ultramicroscopy **110**, 130 (2010)

179. A. Zobelli, A. Gloter, C.P. Ewels, G. Seifert, C. Colliex, Electron knock-on cross section of carbon and boron nitride nanotubes. Phys. Rev. B **75**, 245402 (2007)

180. F. Banhart, P.M. Ajayan, Carbon onions as nanoscopic pressure cells for diamond formation. Nature **382**, 433 (1996)

181. M. Pan, Developing image detectors for in situ TEM applications. J. Chin. Electron Microsc. Soc. **29**, 295 (2010)

Chapter 4
Ultrafast Transmission Electron Microscopy and Electron Diffraction

Alexander Ziegler

Abstract The development and application of pump-probe instrumentation to study rather complex transient events in the solid state, soft matter and life sciences domains has been an area of enormous interest recently. For many years the emphasis has been on the laser-pump and laser-probe approach, followed by the laser-pump and x-ray probe approach. Lately, another method, the laser-pump electron-probe technique, has been gaining interest. Very early experiments using this technique have used electrons to explore gas-phase diffraction of photo-stimulated chemical reactions, followed by experiments in the solid state domain, studying for example, rapid phase transformations, solidification processes, twinning, and shock propagation. The following chapter will focus on the emerging area of ultrafast transmission electron microscopy, or simply ultrafast electron microscopy (UEM), as well as on ultrafast electron diffraction (UED). The topics included are electron sources for pulsed operation and the appropriate emitter materials for this application, electron pulse propagation, electron- and laser-material interactions, the operational challenges during an UEM or UED experiment, i.e., time-resolution and synchronization, the experimental variations, and the challenges encountered when attempting to detect ultrafast experimental processes.

4.1 Introduction

Charged particle beams have, in general, proven to be excellent tools for microanalysis and characterization instrumentation aiming at high-spatial resolution or compositional analysis. These methods usually employ energetic electrons or ions as measuring probes, simply because they are charged particles that can (i) be easily accelerated with an electric field (ii) they can be manipulated and focused to

A. Ziegler (✉)
Microscopy and Microanalysis Unit, The University of the Witwatersrand, 1 Jan Smuts Ave., Johannesburg 2000, South Africa
e-mail: alexander.ziegler@wits.ac.za

A. Ziegler et al. (eds.), *In-situ Materials Characterization*,
Springer Series in Materials Science 193, DOI: 10.1007/978-3-642-45152-2_4,
© Springer-Verlag Berlin Heidelberg 2014

form very fine probes—only a few Angstroms in diameter—by means of an electro-magnetic field, (iii) they interact strongly with the material investigated, and (iv) they can be readily detected.

Taking for example, electrons forming a charged particle beam, one can accelerate these particles via an applied external electric field to almost any desired energy—more precisely a certain velocity. Analytical instruments will usually operate with electron energies between 10^2 and 10^6 V, which translates to electrons travelling at about 2 % and 98 % of the speed of light ($c = 2.998 \times 10^8$ m s^{-1}), respectively. Electrons can reach these speeds because of their very low mass, and typically, accelerating energies in excess of 10^5 V will require relativistic considerations as the velocities start to enter the $0.6c$ domain. At these speeds the key advantage of electrons is their unparalleled small wavelength. At 100, 200 and 300 kV the wavelength attained by electrons is 0.0037, 0.0025, and 0.0019 nm, respectively (relativity considered). Considering the diffraction limited situation with far-field microscopes, $d = \lambda/2$, where d is the spot radius and λ is the wavelength, reducing wavelength is the only means to obtain highest spatial resolution. Thus, the capabilities that fast electrons open up in terms of high spatial resolution characterization are unique. High resolution transmission electron microscopy (HRTEM) has shown remarkable results in the last decade by making individual atoms clearly detectable and visible, even of elements as light as nitrogen and oxygen positioned in a crystal structure. A static high resolution image was (and still is in many instances) revolutionary to the insight and understanding of structure-property relationships in many material systems. Yet, the recent developments toward in-situ high resolution transmission electron microscopy (in-situ HRTEM), enabling the capturing of the time line of a dynamical process in a sample inside of the HRTEM, are opening up even more opportunities to interrogate and begin understanding the nature of dynamic processes like melting, solidification, grain growth, dislocation generation and movement. These instrumental developments are of high interest to many fields of research. The observable processes and the capturing/imaging speeds are on the milliseconds to hours time scale. Since a more detailed description can be found in Chap. 3 in this book, we will not enter this field of in-situ HRTEM characterization any further in this chapter. Instead, this chapter will rather focus on the even faster dynamic processes that occur on a much shorter time scale, namely micro- to femtoseconds. Ultrafast Electron Microscopy (UEM), with special focus on transmission electron microscopy, and Ultrafast Electron Diffraction (UED) will be discussed here. The associated techniques have evolved from the conventional electron microscopy and electron diffraction methods. Yet, their common tool is using the electron as a charged particle that can excellently be manipulated (accelerated, focused and detected) to extract valuable high-resolution information in space and time from the sample and its dynamic process that is underway as we observe. The electron beam thus illuminates a specific area that is undergoing, for example, a rapid structural change; so rapid that it cannot be captured any more with the slower in-situ HRTEM method.

The basic principle of many present-day UEM and UED set ups can be described as follows: A reaction is triggered in the sample and the time-delayed probing beam of electrons creates a distinct signal (diffraction pattern, image) that can be collected at specific time delays relative to the beginning of the triggered reaction. The triggering of a reaction is usually done via pulsed laser irradiation. The idea is simple and plausible, however, the technical implementation, the experimental possibilities, instrumental performance and limitations are multiple and often challenging. A few fundamental changes in the way a conventional electron microscope or electron diffraction set up is operated are mandatory to enable these high time-resolved techniques. The major departure is to convert the conventional, continuous electron particle beam into a very short electron pulse. Other alterations evolve around the timing and synchronization of an UEM or UED experiment and the ultrafast electron detection methods need re-development. The following sections in this chapter will focus on the capabilities and the limitations of these two techniques—UEM and UED, their instrumentation development, experiments and future outlook. Described will be the electron sources and suitable materials for electron guns, electron pulse generation and propagation, laser-material interactions in the sample, the synchronization issues arising in the actual UEM and UED experiment, and finally electron detection in imaging or diffraction mode. A few experimental setups will be presented toward the end of this chapter.

4.2 Electron Sources

4.2.1 Electron Emission

There are three principal methods—and a combination thereof—to extract electrons out of a cathode material: (i) thermionic emission, by heating the electron source material to high temperatures, (ii) field emission, by applying high electric fields, and (iii) photoelectron emission, via irradiation of laser light. All three of these methods find application in the UEM and UED. Especially, a combination of them seems to be the most promising way to achieve high electron densities for very short electron pulses.

In UED and UEM the sources are heated, however, not to the degree that conventional thermionic emission requires. This heating increases the kinetic energy of the electrons inside of the source material, yet not to the extent that they could overcome the binding forces that hold them within the solid structure. The idea is to get them closely to the kinetic energy required to overcome the exit energy—the work function of the material. Conventional electron microscopy cathodes consist of either tungsten (W) or lanthanum hexaboride (LaB_6) tips, which are both widely used for thermionic emission. Tungsten has the advantage of being very robust and having a very high melting point (3410 °C), thus permitting saturation electron current densities of up to 10 A cm^{-2}. However, its work function is comparably high (4.2 eV) and the high operating temperatures cause relatively high vaporization rates, limiting their life time considerably. LaB_6 cathodes instead, with a lower melting point (2210 °C) are operated at lower temperatures, typically last up to 2000 h, exhibit a lower work function (2.66 eV) and the current densities that can be extracted are on the order of 100 A cm^{-2}. The disadvantage of LaB_6 cathodes is their high reactivity at high temperatures, such that the vacuum they are operating in requires pressures less than 10^{-4} Pa.

In field emission the application of very high electric fields to the cathode tip can generate electron current densities of the order of 10^6 A cm^{-2}. Such high electric fields cause a narrow potential barrier at the cathode surface, thus facilitating electron emission by means of the quantum mechanical tunneling effect. The strength of the applied electric fields must generally exceed 10^7 V cm^{-1}, and the cathode needle-shaped tip must be very fine (radii of curvature <1 μm) and oriented in a particular crystal orientation, however, such fields in combination with an insufficient electron gun vacuum will accelerate residual gas pressure ions, that bombard the cathode tip causing detrimental damage to the cathode tip. The required ultra-high vacuum for field emission must therefore be below 10^{-7} Pa.

In photoelectron emission two parameters are important: wavelength and intensity. Irradiating the cathode material with a laser of suitable wavelength provides sufficient energy for the electrons to overcome the work function of the cathode material. For a given material and wavelength of the incident radiation, the number of electrons emitted is proportional to the intensity of the incident light. Hence, irradiating the cathode material with intense laser light will maximize photoelectron emission, and when using nano-, pico-, or femtosecond focused laser pulses, one can generate comparable short electron pulses.

Since the common underlying physical process is to excite bound electrons to the degree where they can overcome the work function and exit the material surface a combination of all three methods can be found in UEM and UED.

The combination of an applied external electric field and the laser irradiation causing photoelectric emission, both promote an initially bound electron to higher energy levels. One has to keep in mind though, that electric field and photoelectron emission are very close, in fact, they overlap when using powerful lasers. Classically, in the photoelectron emission that pertains to weak optical fields the electron's energy is increased by the absorption of a single photon of suitable wavelength and subsequently this photoelectron can exit the material surface. That is the process described above. For strong optical fields, i.e. powerful lasers, one needs to consider the effect of the inherent electric field itself that is part of the optical field. In this case, the (optical) electric field and not the externally applied electric field modifies the emission conditions. The effects are identical: a narrow potential barrier is created at the cathode surface, facilitating electron emission by means of the quantum mechanical tunneling effect. Therefore, a powerful laser irradiating a cathode can be regarded as a photoelectric field emission technique. Tuning or choosing the laser wavelength to match the required energy to overcome the work function via photon absorption will have it act like photoelectron emission. Adjusting and varying the laser intensity and as such its optical electric field carefully, allows to operate a cathode between the classical photoelectron and the electric field emission technique, hence the terms: photoelectric field emission, photo field emission or optical field emission. One can further take advantage of the fact that laser light tends to be linearly polarized and that the direction or angular distribution of the emitted photoelectrons peaks in the direction of the polarization of the incident radiation. Adding the possibility of using a femtosecond laser pulse to excite electrons in a cathode will generate comparable short electron pulses. Generally this combination is very useful and of primary choice to generate short electron pulses for UED. For UEM however, adding the thermionic emission is very practical. Certainly, heating, or partial heating contributes to the increase in the electron's kinetic energy in regard of overcoming the work function of the material, yet, the short laser pulse of suitable wavelength and intensity will actually be controlling the electron emission (electron energy and number of electrons emitted). Operating the setup in quasi-thermionic emission has the advantage that one can easily operate the entire setup with a continuous electron beam by just temporarily blocking the laser pulse and heating the electron source a bit more to reach conventional thermionic emission, and carry out electron optical alignment procedures which tend to be rather difficult in UEM in pulsed mode only.

Due to the physical characteristics of the processes described, photoelectric field emission is prompt with respect to the incident photon. It competes though with the thermally induced electron emission processes, which typically occurs on a time scale of tens of femtoseconds up to picoseconds [1–3]. These events need to be considered when performing the experiment, especially the synchronization between electron emission, the time of flight of the electron pulse over a set distance in the experimental setup, and the time delay desired to capture an event after triggering of a reaction in the sample. The definition of 'time zero' is critical in this case.

In all described scenarios the excited electrons do not leave the cathode tip with uniform kinetic energies. Generally, after emission they exhibit an energy distribution with a full width half maximum (FWHM) of approximately 1, 0.2, and 0.3 eV for thermionic, field, and photoelectron emission respectively. Once the electrons are in vacuum, they can move freely and as such can be accelerated toward the sample.

4.2.2 Emitter Materials

The electron sources, especially the materials for them are usually selected according to a set of requirements that need to be met. The primary goal is to extract as large a current of electrons as possible, because the intensity of the effects that take place in the sample during electron bombardment are greater the more electrons can interact with the sample per unit time. Hence, if the experimental aim is to interrogate a dynamic process on the nanosecond time scale, it needs to be ensured that the electron source can generate as much electrons as

Table 4.1 Brief list of compounds that exhibit acceptable work functions and are suitable as cathode material.

Compound	W_A (eV)	Compound	W_A (eV)
Cs_2O	1.08	ThB_6	2.92
NiCs	1.36	ThO_2	2.96
BaO	1.57	PrB_6	3.12
SbCs	1.84	Y_2O_3	3.3
SrO	2.03	BaB_6	3.45
CaO	2.34	ZrC	3.8
CeB_6	2.59	AgBr	3.9
LaB_6	2.66	AgJ	3.96
SrB_6	2.67	HfC	4
CaB_6	2.86	ZrO_2	4.2

possible in such a short period of time. From conventional electron microscopy, where exposure times during image recording typically range between 0.1 and 5 s, one can deduct that the amounts of electrons required for obtaining a decent, interpretable TEM image is on the order of 10^7 electrons per image, and about 6×10^3 electrons to obtain a decent diffraction pattern (add the numbers of electrons lost at apertures and one easily reaches 10^8 emitted electrons per image). Thus, for a similar image or diffraction pattern to be recorded but at much shorter exposure times, namely nanoseconds or even faster, one requires the same amount of electrons to be present in one single, nano- to femtosecond short electron pulse. Therefore, the set of requirements for ultrafast electron guns involves a low work function of the material, low vapor pressure, high temperature stability, high laser irradiation stability, and the material needs to endure occasional electrical discharges.

Table 4.1 presents a number of elements and alloys that are more or less suitable as cathode material. One can clearly identify Cs_2O as ideal candidate with its work function at 1.08 eV, however, Cs as well as other alkali earth metals is unstable in air, even at room temperature, and when in service, they evaporate at rates that actually cause undesired contamination of the entire cathode housing.

The laser irradiation stability is a very critical parameter. Figure 4.1 demonstrates how pulsed laser irradiation caused severe microstructural damage to a tungsten hair pin cathode that was in service in a quasi-thermionic electron emission set up for UEM. Considering this damage and the invariable material removal—occurring most likely via ion generation and self-bombardment of the cathode when placed inside an accelerating triode set up (cathode, Wehnelt and anode)—it becomes clear how important the material selection, composition and electron source design is. Currently, the cathode designs for UED and UEM are multifaceted. For UED where thermionic operation is generally not required, very sharp metal tips have become the most popular design.

In conventional field emission, electron sources can have emission areas down to the size of single atoms. In the optical field emission variation, however, the source area is typically determined by the area irradiated by the laser—the laser

Fig. 4.1 Laser irradiation induced microstructural damage on a tungsten tip

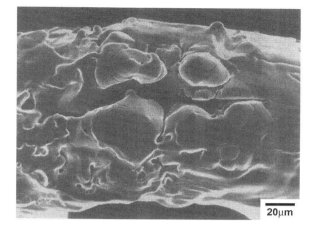

20μm

spot diameter on the cathode, which is generally on the order of one micrometer. Additionally, fine tungsten and gold tips, which are electrochemically etched to very small tip radii (on the order of 20–150 nm) are smaller than the irradiating laser wavelength, and as such provide perfect conditions for an additional electric field enhancement at the tip apex by about a factor of 4–5 [4]. Although these emitter characteristics are desired for very interesting and promising applications, the conditions especially for electric field enhancement will also demand that the material microstructure can withstand such highly localized extracting fields. Generally, single or few-electron emission is achieved, and the electron currents that can be emitted from these fine tips are of the order of pico- to nanoamps. This however, is not quite in line with the demands for other interesting applications of UED or UEM. For these, much higher electron currents are required—on the order of a few hundred microamps. Hence, the small material volume at the fine tips, which is exposed to these demanding operating conditions, has obviously less capacities for increased energy deposition and dissipation, and the diffusion of excess electron–gas heat than their bulk counterparts. One has to consider that intense laser-material interactions are rather used to study purposeful material damage, i.e., inflicting very steep temperature gradients, phase explosions, extremely rapid and highly localized electron–gas heating, electron–phonon scattering, shock wave propagation, pair excitation, etc. One should therefore not expect that in delicate cathodes these detrimental interactions between intense laser light and material are suddenly not taking place.

Comparably "bulk" cathode designs are currently being used for UEM applications. The current research focus is more on the material compatibility and resistance to the demanding operating conditions. Hair pin type cathodes, typically made of fine tungsten wires (50 μm in diameter) with a suitable coating at the tip have been experimented with. Table 4.2 lists a few material compositions that are possible, yet their reliability and especially their reproducibility is limited.

Table 4.2 Brief list of combinations of coatings and carrier materials that exhibit acceptable work functions and are suitable for pulsed cathode operation.

Coating	Carrier	W_A [eV]	Coating	Carrier	W_A [eV]
Cs	Pt	1.39	Na	Pt	2.1
Cs	W-oxide	1.44	BaO	W	2.1
Rb	Pt	1.57	La	W	2.13
K	Pt	1.62	Ce	W	2.71
Cs	W	1.64	U	W	2.95
B	W	1.75	Zr	W	3.14
CaO	Ni	1.77	Th	W	3.2
Ba	PtIr	1.77	W-oxide	W	6.24
Ba	W-oxide	1.8	Ni-oxide	Ni	6.34
Na	W	2	Pt-oxide	Pt	6.55

The coating tends to be either molten or sintered to bond to the hair pin tip. Molten coatings exhibit often very smooth surfaces, occasionally with small elevations and bumps. The latter, however, instead of becoming preferred electron emission locations due to high electric field concentrations, they are most often preferred sites for electric discharge generation at high extraction and acceleration fields. Moreover, the molten material relocates away from the actual tip of the hair pin due to capillary forces driving the liquid up the hair pin shaft, and once in the liquid state, phase separation and/or alloy formation with the tungsten wire can take place. The aim is to not have the tungsten come through the molten layer, as its high work function is not desirable and it should thus not be exposed to the laser irradiation. Alternatively, the sintered material at the hair pin tip tends to exhibit rather rough surfaces, but bonds well to the tungsten wire and remains in place at the tip, and does not undergo alloy formation. The chances of electric discharges are given, but because the surface roughness covers the entire tip and is not just sparsely distributed over the entire surface area, electric discharge occurs less often than with the molten counterparts. In fact, the surface roughness creates a much larger surface area than the smooth molten kind of tip. This proves to be of great advantage when irradiated by the laser, because many more electrons can be emitted simultaneously. In projection, the irradiated area corresponds to the laser spot on a smooth surface, yet the true surface area that is excited and where photons deposit their energy can be much larger on a rough surface. The laser photons penetrate the rough surface at all locations within the laser spot. However, some locations are reached and excited earlier than others and, unlike on a smooth surface, many locations are not perpendicular to the incident irradiation. The consequence is that electron emission is most irregular and not perfectly in phase with each other. Moreover, the electrons tend to be emitted in all possible directions and are not necessarily pointed toward the anode or in the direction of the laser polarization. Hence, the number of emitted electrons is large, but their orientation and energy distribution is large too.

These sintered cathode tips are most promising in terms of high electron yield per laser pulse. A few carefully selected and manufactured material combinations have shown electron emission currents up to 1.5 milliamps. Especially ZrC coated

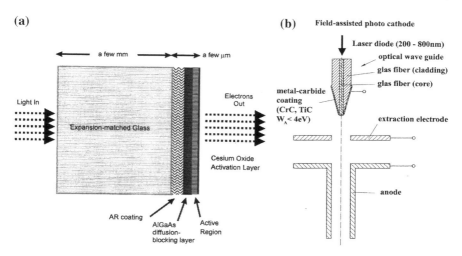

Fig. 4.2 **a** Negative electron affinity photocathode, using back-illuminated Cs_2O films on semiconductors [5], **b** Fiber optic wave guide coated with suitable material for electron emission, back-illuminated through the wave guide with a laser to emit electrons at the tip [6]

tips fall into this category, and they have also proven to have acceptable lifetimes (weeks to months). Their emission and operation characteristics (work function, resistance to laser induced damage, temperature stability, low reactivity, good bonding to carrier wire) are satisfactory.

There are a few other cathode designs that are being used in UED and UEM setups. One of them is to back-illuminate a thin metal foil with the pulsed laser to emit electrons out from the front. A development relying on the same principle is using negative electron affinity photo cathodes, Fig. 4.2a. Here the laser enters through an expansion matched glass covered with an anti-reflection (AR) coating, a AlGaAs diffusion blocking layer, an active region, and finally the Cs_2O activation layer that emits the electrons. The advantages are: low noise, low energy spread, high brightness, and a submicron emission area. The greatest disadvantage is the continuous Cs desorption and diffusion, requiring almost continuous cesiation of the activation layer. The consequences are lifetimes of only 75–100 h at 50–160 nanoamp emission currents. Another interesting setup is to coat a short (few centimeters) fiber optic guide with a suitable material compound that yields high density electron pulses when back-illuminated with a powerful laser pulse through the fiber optic guide, Fig. 4.2b. The problem here is the mediocre bonding characteristics of the coating materials to the glass fiber optic, laser irradiation endurance and possible absorption, damage and ensuing loss of power inside the wave-guide. Moreover, the coating has to be thin enough, such that it is almost light transparent, for the photons coming from the backside to still generate electron emission out of the front.

4.3 Electron Pulse Propagation

Once the electrons are emitted and in free space they are accelerated and focused onto the sample. For UEM, currently the same gun designs that are most common for conventional electron microscopes are in use. That is, the emitting cathode is opposite a centrosymmetric set of electrodes, the Wehnelt electrode and the anode. The cathode is at a high negative potential, whereas the Wehnelt electrode is only slightly negative with respect to the cathode potential. This creates a potential threshold, causing an accumulation of electrons and a localized space charge right in front of the cathode that the emitted electrons have to overcome to continue acceleration toward the anode. By varying the Wehnelt voltage one can control this potential threshold and as such manipulate the emitted electron current within a wide range. The Wehnelt electrode and the anode also function as an electron optical lens system that shapes and directs the electron pulse—a necessary function. Generally, this electrode combination cathode, Wehnelt and anode allow forming a small disc-shaped image of the emitting area on the cathode, a crossover, and as such generating a powerful electron probe where a large current density and a small angle of divergence can be achieved. This however, can only be accomplished, when the Wehnelt electrode potential creates a threshold that is so high that only the most energetic emitted electrons that are also directed straight toward the anode can make it through the space charged area in front of the cathode. This automatically filters out low energy electrons and those with a high angular distribution. One certainly does not want to reduce the eagerly generated maximum number of electrons in a single pulse (up to 10^{12} electrons/m^2/ps), yet to form a usefully shaped electron pulse in terms of energy and velocity distribution, one needs to carefully strike a balance between maximum number of electrons and suitable electrons.

As much as electrons can be influenced by an external electric and/or magnetic field, their inherent charged state can become an insurmountable obstacle when it comes to highly dense electron pulses in UED and UEM. The biggest challenge is the repulsive Coulomb force acting on the individual charged electrons when in close proximity—space charging. It limits the formation and propagation of ultra-short (femto- to nanosecond) electron pulses considerably [7–10]. This detrimental condition is certainly met at focal points and crossovers along the beam path in UEM. Ultrashort electron pulses broaden in space and lengthen in time by orders of magnitude depending on the distance (or time) traveled and the amount of electrons they consist of. For example, a 200 fs long, 60 kV electron pulse containing 10^{10} electrons/m^2 quickly elongates to 2 ps if traveling for about 5 ns. To avoid pulse lengthening it may not travel for longer than 0.5 ns (Fig. 4.3), which corresponds to 7.2 cm. The lateral spread in space of an "overcharged" electron pulse causes similar problems. Although electrons can be focused to spots only a few Angstroms in diameter, this does not apply anymore when more than several millions of electrons need to go simultaneously through focal points and crossovers (only nano- to micrometers in size) as required in electron microscopes. Besides

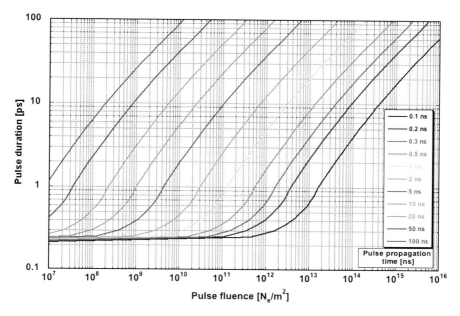

Fig. 4.3 Electron pulse spread depends on the number of electrons it contains and the time (or distance) it travels after being emitted from the source

the space-charge related effect of pulse broadening, the energy spread ΔE of the accelerated electrons (30–300 keV) will cause similar dispersion of the electron pulses. Electrons with slightly more energy are accelerated to faster speeds and will travel further than less accelerated electrons within the same electron pulse—the pulse begins to stretch. These implications mandate short distances when employing ultrashort electron pulses, which is inherently a problem in relatively large electron microscopes. Already, one single electro magnetic lens may exceed the maximum distance an ultrashort electron pulse may travel before too much broadening and lengthening occurs.

Hence, pulse compression techniques are required. These methods range from simple reflection of the dispersing electron pulse off flat electrostatic fields, whereby the faster electrons penetrate the field deeper and reverse direction of travel later than the slower electrons, to radio-frequency (RF) cavities [10], gradient acceleration and traveling self compression, and correlations between a synthetic optical-field and the electron-pulse, employing the effects of the ponderomotive force [11], (Fig. 4.4). Experimental setups for UED allow meeting such requirements best, since distances can be kept short and modifications to the setup can be easily implemented—from the electron source, over pulse compression elements, to the sample and to the detector. It is important to keep in mind that the electron pulse will always be subject to Coulomb forces broadening it in all dimensions. This certainly also applies after the pulse exits any pulse compression section(s) installed along its path—it will invariably start broadening again. Pulse compression can

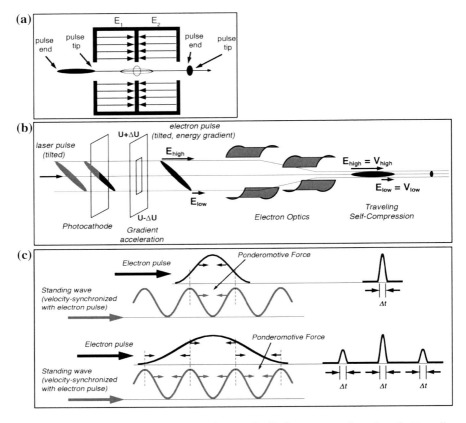

Fig. 4.4 Electron pulse compression techniques: **a** Radio frequency-cavity, where fast traveling electrons at the pulse tip are retarded whereas slow traveling electrons at the pulse end are accelerated. **b** Tilted electron pulse compression, where an accelerating potential gradient (U ± ΔU) causes an energy gradient, inducing a controlled velocity gradient and facilitating traveling self-compression of the electron pulse. **c** Ponderomotive force based pulse compression, where a synthesized standing optical wave is velocity synchronized with the electron pulse and the ponderomotive force is able to shape the incoming electron pulse to extremely short electron pulses. Electron pulse duration is indicated by Δt, where t is time

only temporarily re-shape the electron pulse at specific positions. Thus, it is helpful to perform pulse compression right before the sample, where pulse length is most critical for the achievable time resolution of the instrumentation.

Single-electron pulses instead do not experience Coulomb repulsion. They can be regarded as the ultimate non-dispersing electron pulse. They are only subject to the inherent energy spread ΔE and its consequences when accelerated. They can be employed in UEM, with all the bulky electromagnetic lenses, and they can be employed for UED. The drawback of single-electron pulses is that single-electron scattering events are insufficient to capture the entire ultrafast action occurring in the sample all at once. A solution to that is to increase their repetition rate. But even with very high repetition rates, one needs to revert to a recurring reaction in the sample, that is, the triggered reaction in the sample needs to be reversible [12–14]. A one-time, irreversible reaction is impossible to track with just one electron. A repetitive reaction in the sample can be examined only because every time the reaction is initiated a single-electron scattering (probing) event is recorded, contributing each time to the completion of the final picture, either in diffraction or in imaging mode, UED or UEM respectively. At repetition rates of MHz or GHz, the entire experiment could be concluded in less than a second (provided the reaction in the smple returns to the original state that fast), with sufficient scattering events and a useful diffraction pattern that can be evaluated. A multitude of reactions in the chemistry and biology domain are ideal candidates, as well as reversible solid–solid phase transformations and magnetic domain switching in the materials sciences. However, phase-explosions, melting and re-crystallization events are most likely not.

There is currently no better option for these irreversible reactions in a sample than using multi-electron pulses, containing as many electrons as can be generated and also compressed into one pulse [15–18]. Emission limits are currently $\sim10^{12}$ electrons/m^2 per sub-500 fs pulse. One has therefore only one single shot with one single multi-electron pulse to perform the entire experiment, because after that the sample has either transformed irreversibly or has been destroyed. There are cases and experiments where the reaction is not as fast and femtosecond pulses are not necessarily required to interrogate the time line of the process. In such cases one can indeed revert to multi-electron pulses, where space-charge effects and pulse broadening is not as critical, because the reaction time line is on the order of nanoseconds or even microseconds [17–25]. At these speeds—taking space-charge effects into account—one can perform time-resolved electron microscopy, UEM. The high spatial resolutions as attained with non-pulsed electron microscopes cannot be achieved though, because the correlation between space and time resolutions involves some inherent uncertainties Δt and Δx at the ultrafast time scale [26].

4.4 Electron- and Laser-Material Interactions

The simple concept behind ultrafast experiments was summarized at the beginning of this chapter: A reaction is triggered in the sample and the time-delayed probing beam of electrons creates a distinct signal (diffraction pattern, image) that can be collected at specific time delays relative to the beginning of the triggered reaction. In the current scientific literature the word "ultrashort" refers to the time domain ranging from femtoseconds to picoseconds. Accordingly, the word "ultrafast" refers to physical, chemical, and biological events that occur on that time scale, and hence the terms femtophysics, femtochemistry and femtobiology can be found in the recent scientific literature. Femto materials science seems to be less known, possibly because it overlaps with femtophysics and in some aspects with femtochemistry, because ultimately we are all trying to understand how atoms and electrons behave. Even faster processes than "femto" occur on the attosecond time scale. The main driving force for ultrafast science in physics, chemistry, biology and the materials sciences is the development of better, faster, brighter and more stable lasers. Without pico-, femto- and now attosecond laser pulses ultrafast science would hardly be possible.

4.4.1 Ultrafast Conditions in Matter

The physical situation in a solid when it interacts with an ultrashort laser pulse can generally be described as follows (the time scale of such process stretches over several orders of magnitude): The laser pulse impinges on the surface of a material (assume a metal), and immediately after photon absorption via inverse Bremsstrahlung the free electrons have a non-thermal energy distribution and therefore no specific temperature can be assigned to the system. This condition changes within the following few hundred femtoseconds as the electron gas thermalizes due to electron–electron scattering. A precise, but extremely high temperature (several thousand Kelvin) can be attributed to the system now. The ensuing transfer of energy to the lattice requires a few picoseconds, due to electron–phonon scattering. Many electron–phonon collisions are necessary—phonons are much heavier than electrons ($m_p/m_e \sim 10^5$)—in order to move phonons and to heat up the lattice. The lattice thermalizes and approaches an equilibrium temperature afterwards via phonon–phonon scattering. At sufficiently strong peak intensities material can be ablated.

The electron–electron and the ensuing electron–phonon scattering process has traditionally been described by a two-temperature model [27, 28], including two separate heat capacities, one belonging to the excited free electron gas (T_e and C_e), the other to the lattice phonons (T_l and C_l).

$$C_e \cdot \frac{\partial T_e}{\partial t} = \Delta(k_e T_e) - g(T_e - T_l) + E(z, t) \tag{4.1}$$

$$C_l \cdot \frac{\partial T_l}{\partial t} = \Delta(k_l T_l) - g(T_e - T_l) \tag{4.2}$$

where E is the absorbed laser energy, g the electron–phonon coupling constant, C_e and C_l the heat capacity, and k_e and k_l the thermal conductivity of the excited electron gas and the lattice, respectively. C_e is approximately two orders of magnitude smaller than C_l, and hence the increase in temperature of the bulk will be only a small fraction of the excited free electron gas temperature. However, this description is valid only when T_e is known. Unfortunately, this is not the case at times shorter than 100 fs—before the electron gas thermalization. For shorter times, a model based on effective potentials needs to be used [29].

The time domain shorter than 100 fs is scientifically becoming very interesting, and the precise absorption and electron-relaxation processes occurring in the very localized volume at these extremely short times are not fully understood yet.

Extremely high temperatures and pressures can be expected. The lattice may instantly disintegrate under such conditions, because a transfer of energy from the excited electrons to the lattice (electron–phonon scattering) may not be possible. The result is a direct, ultrafast solid–liquid (and/or solid-plasma) phase transformation [30–32].

Under less than critical conditions but immediately after electron relaxation, the effective potential instantly takes on its new condition with a new equilibrium separation of the ions. However, the rather sluggish response of the much heavier ions to the altered effective potential, which is forcing the ions into new positions, results in enormous transient pressures building up in the irradiated area. The result are extremely high pressure gradients that can either disrupt the local lattice and/or send shock waves (compressive and tensile stresses) into the surrounding microstructure, causing spallation at free surfaces and interfaces intercepting the shock wave path. The interface bonding characteristics, especially those of thin films deposited onto substrates could be examined.

The circumstances are different during the energy deposition of a nanosecond long laser pulse. When the deposited laser energy is sufficient and the time frame long enough to establish a relatively stable temperature distribution in the sample, slightly above the melting temperature, then the sample, for example a thin film, has reached a highly local hydrodynamic state of imbalance [26]. Capillary forces at the surface(s) are acting and will result in shear stresses—in accordance with the dependence of the surface tension on temperature and volume fraction of surface impurities (e.g., oxygen), $\gamma = \gamma(T, X)$.

$$\frac{\partial \gamma(T,X)}{\partial r} = \left(\frac{\partial \gamma}{\partial T}\right)_X \frac{\partial T}{\partial r} + \left(\frac{\partial \gamma}{\partial X}\right)_T \frac{\partial X}{\partial r} = \left(\frac{\partial \gamma}{\partial r}\right)_X + \left(\frac{\partial \gamma}{\partial r}\right)_T \quad (4.3)$$

Impurities will modify the surface structure by replacing lattice atoms, thus altering the surface tension characteristics. The resulting shear stresses will initiate a transport of liquid (melt) (Marangoni effect). The thermal gradient $\partial T/\partial r$ and the compositional gradient $\partial X/\partial r$ create a thermo- $(\partial\gamma/\partial r)_X$ and/or a chemo-capillary $(\partial\gamma/\partial r)_T$ driven flow of matter. Accordingly, three scenarios can be distinguished in Fig. 4.5a, b and c.

In case (a), pure metals, there are no contaminating elements involved ($X = 0$), and thus thermo-capillary forces will control the shear stresses. The Gaussian shaped energy distribution of the laser pulse imposes a negative thermal gradient $\partial T/\partial r$ onto the thin film, whereas the thermal coefficient $(\partial\gamma/\partial T)_{X=0}$ is negative for pure metals. As a result, the shear stresses $\partial\gamma/\partial r$ are positive, pushing the molten material unidirectionally toward the colder, solid perimeter of the irradiated area. Under such conditions, a hole in the film can be created at the center of the melt. In case (b), contaminated metals, $X \neq 0$. Before, and at the beginning stages of irradiation, there is no, or no significant, gradient of the contaminating species, $\partial X/\partial r = 0$. Contaminants are homogeneously distributed. The initial flow of material will be controlled, as in case (a) only by the thermo-capillary forces, moving outward (radial). As the energy deposition of the laser continues to heat up the thin film, a low-viscosity surface layer of melt tends to flow to the colder perimeter faster than the "bulk" material inside the film. This creates a redistribution of the

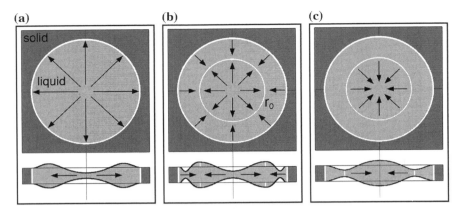

Fig. 4.5 Laser-material interactions on thin metal films. **a** Pure metals, **b** Contaminated metals, and **c** Over-heated thin films

impurities/contaminants (impurities are moving away from the center of the irradiated area) and thus $\partial X/\partial r > 0$. The growing chemo-capillary force starts to counter balance the thermocapillary force, and the initial outward flow of liquid matter may now be retarded (and even reversed). The governing parameter is the degree of surface-active contamination. The surface tension and its temperature dependence $\partial\gamma/\partial T$ are strongly affected by the overall impurity concentration at the film's surface. If $X \gg 0$, then $\gamma(T)$ exhibits a local maximum at a specific temperature T_0, and the thermal coefficient $(\partial\gamma/\partial T)_X$ is positive below and negative above T_0, respectively. The laser fluence (or the deposited energy) becomes now a controlling parameter. If the temperature in the entire irradiated area remains below T_0, and with $\partial T/\partial r < 0$, the thermo- and the chemo-capillary forces will move the melt toward the center. If $T > T_0$ within a certain range $r < r_0$, then two competing driving forces push the liquid in opposite directions. For $0 < r < r_0$, the thermal coefficient $(\partial\gamma/\partial T)_X$ is negative, and the liquid flow is in many cases driven outward by the thermo-capillary force (the thermal gradient tends to be steeper than the compositional gradient, $(\partial T/\partial r) > (\partial X/\partial r)$). Outside this area, $r > r_0$, the temperature is below T_0 and $(\partial\gamma/\partial T)_X$ is positive, and concomitantly the flow of liquid directed toward the center. This scenario can result in a ring (or network) of liquid material suspended temporarily in space before it breaks apart. Depending on the dominating movement (in or outward), the ring/network or fractions thereof may impact on the perimeter of the created hole or collapse in the center. Those dynamics can be highly complex. In case (c), over-heated thin films, a very high temperature at the center of the irradiated area causes the surface-active impurities to vaporize right away. This results rapidly in $\partial X/\partial r > 0$ and in addition, since the central area affected by this mechanism is relatively small, one finds that the temperature gradient for small radii $\partial T/\partial r$ here is not as steep, and therefore $(\partial T/\partial r) < (\partial X/\partial r)$. In this case, the chemo-capillary forces dominate, driving most of the liquid toward the center and detaching it from the perimeter.

Laser pulses with high fluences can catapult a material system into regimes of extreme pressures and temperatures never attained before [33], and this with very high heating and cooling rates (up to 10^{11} K s^{-1}). The interactions are predominantly non-linear. The internal pressure generating mechanisms are different at different times of observation—femto- to nanoseconds, electrons and lattice ions respectively. Upon quenching, the material system may follow previously unknown paths of nucleation, crystallization, and grain growth. As a result, microstructures may also exhibit extraordinary material properties as in the case of "black silicon"—a surface modification of silicon, resulting in very low reflectivity after being irradiated by a set of laser pulses in a halogen gas environment [34–41]. The surface of such laser-treated silicon develops a rather self-organized structure of micrometer-sized cones and the concomitant "new" material properties include enhanced absorption, extending into the infrared spectral range below the band gap of silicon. Pulsed laser-material interactions and the results are multifaceted: solids are able to undergo a phase transition without appreciable heating of the bulk lattice, electron dynamics can be triggered explicitly at solid interfaces, demagnetization and a dielectric breakdown in solids can be induced [30, 42], and rapid phase changes can be set off locally in memory alloys and small structures. Examination of pulsed laser-treated micro- and especially nanostructures on surfaces and their already size-dependent properties is as appealing as the possibilities that open up when a laser pulse is tightly focused onto a plane inside of a transparent material [31, 32, 43–51]. Nonlinear optical absorption occurs, leading to very high temperatures and a change of the material microstructure at the focal point. This type of tailoring (or machining) of the material very locally has had some significant implications in medical applications. Corneal tissue can now be removed without any noticeable indication of thermal damage to the neighboring tissue, and functional neurosurgery could possibly also benefit from the highly localized ablation possibilities of ultrafast laser pulses. Furthermore, the extremely high intensities of such short laser pulses can cause a plethora of laser-material interactions [33], including laser-induced collision through pair excitation, double ionization, interatomic coulombic decay, stimulated scattering, soft bonding and vibrational trapping in molecules, and optically excite non-thermal electrons in nanostructures. A few examples of the outcome of laser-material interactions and the material response are shown in Fig. 4.6. Laser induced quenching is also able to "freeze" a microstructure into a semi-stable state not attainable otherwise, involving for example high dislocation densities. This opens the possibility for precise strain engineering: a purposeful, precise placement of residual strain fields and strain generating areas of high dislocation density, which is currently a method under consideration to further reduce the dimensions of the next generation of electronic devices [52]. Precisely positioned residual internal as well as surface strain fields can strongly affect the free carrier mobilities in semiconducting materials.

One needs to be aware of the fact that in UED and UEM only very thin samples can be investigated and that the results may not always be representative and applicable to bulk properties. For conventional TEM investigations, also on thin

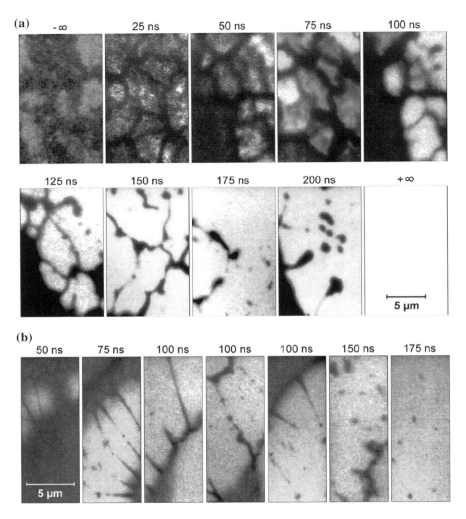

Fig. 4.6 UEM bright-field images of: **a** Creation and disintegration of a network of melt in a laser-irradiated 45 nm thick Ag foil. Surface evaporation and an ensuing decomposition of the cell walls (after 100 ns) accompany the disintegration of the network. **b** The high viscosity of the melt of a laser-treated 33 nm thick Ge foil helps explaining the formation of very long and thin filaments of melt. They rupture after 75–100 ns due to a Rayleigh-instability in melt droplets (150–175 ns). **c** Phase explosion in a 30 nm thick Ni foil. The untreated sample exhibiting bend lines ($-\infty$), melting (0 ns), abrupt generation of gas bubbles (5–13 ns), and expansion and compression waves (15–30 ns), final state $+\infty$. **d** UED diffraction pattern illustrating the decrease of the lattice spacing from Ti {10–11} hcp to{110} bcc due to heating. (*top*) Pattern of the Ti film before treatment (*bottom*). Displacement of the reflections framed in (*top*) with increasing time after the laser pulse. Each picture was taken at the indicated time after the treating laser pulse from a fresh location of the same film, treated with a pulse of the same fluence. (images **a**, **b**, **c**, **d**: courtesy Dr. Holger Dörner [25])

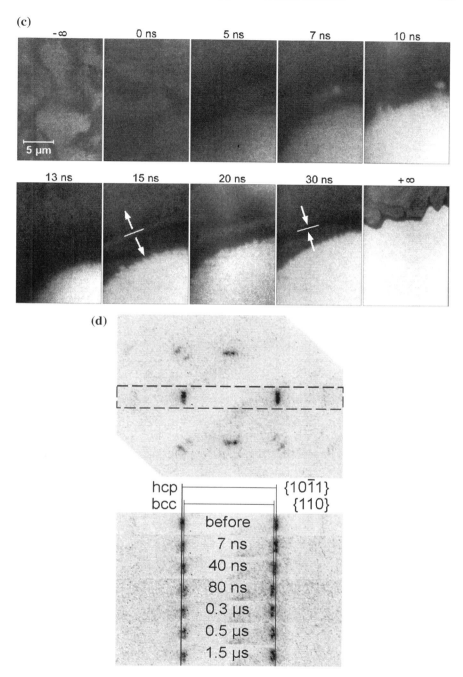

Fig. 4.6   continued

samples, this issue has been resolved in the past by careful and repeated examination and consideration of such boundary condition effects. For the dynamic aspects of thin samples this may be different. The presence of surfaces on both sides and the fact that the laser that usually triggers a reaction of interest in the sample can fully penetrate and activate the entire sample volume (laser penetration depth of approximately 100 nm), certainly represents a different scenario than the same laser triggering a surface reaction on a bulk sample, where no second (back or exit) surface is reached, which may influence the reaction.

Nevertheless, in order to comprehend the dynamic processes behind such interesting physical and chemical changes one also needs to start thinking about the time domain between micro- and femtoseconds. Electrons, atoms, molecules, and even larger living subjects like cells tend to move, change shape, position and conformation, and possibly also composition over time—short times. There seem to exist fundamental event durations that will need to be taken into account when addressing ultrafast science aspects. For example, atomic motions at the length scale of a few Angstroms take place within a few femtoseconds. Localized processes involving optical phonons, defect formation and non-thermal melting occur in the femtosecond to picosecond, time range. The ensuing events of acoustic phonon generation, structural phase transitions and even strain propagation occur at much longer times, from nano- to microseconds. The hierarchy of length scales may therefore correlate with time scales. Common to all courses of action is that the more large-scale "global" events, occurring at later times have their initiation at the very short times and distances. These time scales are becoming very critical when designing a UED or UEM experiment. It will be almost impossible to capture an event occurring within 10 fs with a 1 nanosecond long electron pulse, even if the event is timed such that the electron pulse and the reaction of interest perfectly overlap in time. The number of electrons in the entire electron pulse probing the sample is set to, for example 10^6, but since the reaction is 10^5 times faster/shorter, only 10 out of the 10^6 electrons will actually be probing the reaction—clearly a condition yielding a very poor signal-to-noise ratio. Experimentally, one must consider that it takes a few picoseconds to generate a decent electron pulse and this pulse needs to travel for a few hundred more picoseconds from the source to the sample. Hence, the synchronization between the pulse's time scale and the triggering of the reaction in the sample becomes an important experimental parameter. Timing is key.

4.5 Time Resolution and Synchronization

4.5.1 Electron Pulse Profile

The best achievable time resolution and the optimum synchronization of events in the entire UED or UEM experiment is fundamentally tied to the duration of the probing electron pulse. Currently, it is the primary factor limiting temporal resolution, and it is therefore very important to properly characterize and determine pulse length and shape. However, this is not a straight forward process and it is furthermore complicated by the fact that the electron pulse broadens due to Coulomb repulsion acting between charged particles. Any pulse length and shape characterization can therefore be regarded only as a specific snapshot in time and space of the traveling electron pulse. At any other location the electron pulse will have a different length and shape. It is therefore most appropriate to characterize the electron pulse at the location where it matters most—the sample, where the reaction occurs. A short and well defined pulse at this location will determine the time resolution of the UED or UEM experiment.

Currently a good technique that allows electron pulse shape characterization is a streak camera. Here the temporal profile of the electron pulse is transformed into a spatial profile and recorded on a detector, by causing a time-varying deflection of the incident electron pulse. For electrons, a quickly modulated electric field between a pair of deflection plates serves this purpose and the electron pulse is swept over a linear detector, usually a charge-coupled device (CCD). The resulting image, a "streak" obtains detailed information about the pulse duration and other temporal properties of the pulse. The time resolution of such a streak camera is related to its response to the changing electric field at the deflector plates, the dimensions of the deflector assembly and its distance to the detector. Streak speeds of up to 2×10^8 m s^{-1} have been achieved [14, 53, 54] resulting in an instrumental time resolution of approximately 500 fs. However, the physical dimensions of these streak cameras (several centimeters) are also an obstacle to the achievable precision too, because the electron pulse will continue broadening inside the detection setup. As seen earlier, a 200 fs long electron pulse containing 10^{10} electrons/m^2 quickly elongates to 2 ps if traveling for about 5 ns. To avoid pulse lengthening it may not travel for longer than 0.5 ns, corresponding to 7.2 cm.

A different approach to pulse shape measurement relies on the optical autocorrelation, a technique well known from optical femtosecond laser pulse metrology [55, 56], Fig. 4.7. Here, successive optical intensity autocorrelations of increasing order are employed to measure the exact pulse duration. The setup is identical to a Michelson Interferometer, where the incident laser pulse (the one that initially generates the electron pulse from a cathode) is split and reflected by mirrors that can be set to introduce a variable time delay of a few picoseconds. Thus, two consecutive laser pulses arrive at the cathode, just a few picoseconds apart, generate two (ideally) identical electron pulses that are then focused via an electromagnetic (or simply a magnetic) lens onto the sample plane. The sample plane is where the pulse length measurement should take place as otherwise the pulse will change shape and length along a continued path. The fact that the focusing action compresses the electrons into a smaller volume, invokes the Coulomb repulsion among the electrons to react against this focusing action. Temporal and spatial broadening of the two electron pulses causes them to begin overlapping and interacting with each other depending on their electron density distribution. This represents an equivalent to the non-linear optical medium required for the optical autocorrelation of femtosecond laser pulses, where interaction depends on intensity. The resulting beam profile is detected with an electron detector. Assuming a Gaussian

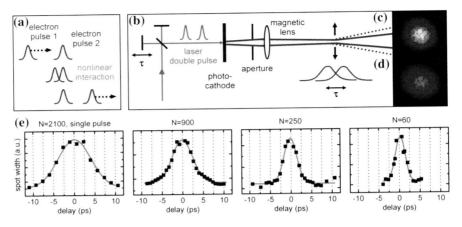

Fig. 4.7 Electron pulse autocorrelation. **a** Concept of an autocorrelation measurement with electron pulses exploiting the nonlinearity in free space. **b** Experimental arrangement. A laser double-pulse generates two identical electron pulses with a delay. A magnetic lens system focuses the electron pulses toward a narrow spatial waist, where Coulomb repulsion leads to a spatial beam broadening that depends on delay time. **c** and **d** Beam profiles are shown without (**c**) and with (**d**) temporal overlap between the two electron pulses. **e** The traces show the measured spatial beam diameter versus the delay between the electron pulses for various numbers of electrons. The green traces are fits to Gaussian functions. Note the decrease in width with the decrease in the number N of electrons; here, N is relative (courtesy Dr. Peter Baum [56])

time profile of the electron pulses, the autocorrelation width is $\sqrt{2}$ longer than the actual pulse width (length), and as such a rather precise temporal pulse characterization is possible at the sample plane.

4.5.2 Synchronization at the Sample and Time Zero

In UED and UEM the probing electron pulse and the laser pulse that triggers a reaction in the sample are propagating along different paths. In some setups they might approach the sample in an almost collinear form, while in others they meet almost perpendicular to each other. The location of interest is the sample, the point of coincidence, where both, the electron and the laser pulse should arrive simultaneously or in slight relative delay (the electron pulse probes an ongoing reaction set off by the laser pulse shortly before). Ideally, both beam paths should be collinear, but a problem arises when the angle between both beam paths increases due to experimental and/or geometric setup conditions: their discrepancy in propagation velocity and their point(s) of coincidence. The laser travels at the speed of light, **c**, whereas the electron beam travels at a speed according to the acceleration potential (or energy). As pointed out initially, at about 100 kV acceleration the electron reaches 0.62c—a significant difference in velocities with which both pulses impinge on and travel through the sample. If the UED

Fig. 4.8 A 10 fs electron pulse and a 10 fs laser pulse (35° relative angle) impinge on a 50 nm thick sample. The footprints of both pulses are set to match. The laser pulse travels at speed of c, whereas the electron pulse travels at speed of 0.62c. "Time zero" can be set to t_0, t_1 or t_2, each giving a different resulting image or diffraction pattern

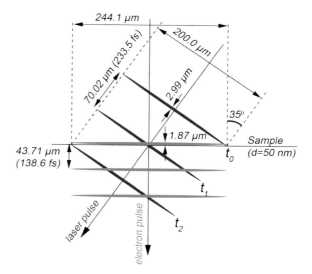

and UEM setups are operating in transmission mode, that is, the sample is electron transparent, then the penetration depth of the laser pulse (approx. 100 nm) usually exceeds the sample thickness (10–100 nm, depending on the sample material). Assuming no absorption or scattering effects take place, the front of the laser pulse traverses or reaches this distance (100 nm) in 0.33 fs. The front of the electron pulse traverses the same distance in 0.53 fs. Considering the instrumental time resolution currently attainable, there is no question this is much faster. This is the case when both pulses (electron and laser) are collinear. Due to experimental geometric conditions one cannot always attain collinearity and both pulses reach the sample at a relative angle. This condition poses a serious problem when attempting to reach highest temporal and spatial resolution. An example scenario could be the following (see Fig. 4.8): A UED or UEM experiment is set for using electron pulses (duration: 10 fs, acceleration: 100 kV, speed in vacuum: 1.875×10^8 m s^{-1} (=0,625c), pulse focused to: 244.1 μm diameter, direction of incidence relative to the sample normal: parallel) and laser pulses (duration: 10 femtoseconds, speed in vacuum: **c** (=2.998 × 10^8 m s^{-1}), pulse focused to an oval shape: 200.0 × 244.1 μm, direction of incidence relative to the sample normal: 35°). The oval shaped laser pulse is required to match the illuminated area (footprint) of the electron pulse. The 200 μm axis will, in projection at 35°, be 244 μm in size. It will be much easier to shape the laser pulse to such an oval form, instead of the electron pulse. Assumptions: the experiment is set to using one electron and one laser pulse to perform the experiment (call it single-shot experiment here, although the geometric considerations apply to repetitive, multi-pulse experiments as well), the laser and the electron pulse exhibit a Gaussian intensity profile, yet only the areas of illumination will be considered here, laser energy deposition will be considered as equal and homogeneous over the entire illuminated area, the repulsive Coulomb forces between the electrons are not posing a problem or pulse compression was successful up to the sample plane, and the sample will be characterized in transmission and it is therefore electron transparent with an overall thickness of 50 nm (a self-supporting foil).

These boundary conditions imply the following. The physical, spatial dimensions of the laser pulse are $200.0 \times 244.1 \times 2.99$ μm, in essence a flat disc approaching the sample at $35°$. The electron pulse, since it is travelling at a lower speed, has spatial dimensions 244.1 (diameter) $\times$ 1.87 μm. Note, the "thickness" of both these discs is much larger than the sample thickness of 50 nm, hence the electron pulse, coming in perpendicular, will easily cover the entire sample thickness, front to back, as it propagates through it. The laser pulse instead, will not be able to do this under these conditions. It will only be able to do this when incident perpendicular (in current existing instrumentation rarely the case, see Fig. 4.9). In this example, the first laser energy deposition will be considered "time zero", $t_0 = 0$, that is, when the laser pulse first touches the sample on its way through. The time elapsed until the laser pulse is half way through the sample is considered $t_1 = 233.5$ fs (distance traveled $= 70.02$ μm). Consequently, the time needed for the laser pulse depositing its last quantum of energy when exiting the sample is $t_2 = 467$ fs (distance traveled $= 140.04$ μm).

In the following experiment we want to record the "time zero" event (experimental setup as depicted in Fig. 4.8). The laser pulse is just depositing its first energy quantum into the sample and the electron pulse will capture that and a few following moments, hence, the location of the front end of the electron pulse at "time zero" is exactly at the entry surface of the sample. Until the entire electron pulse exits the sample, that is, the back end of the electron pulse exits the sample's bottom surface 10.26 fs will have passed. During those 10.26 fs, the laser will have been able to deposit more energy into the sample and progressed 3.07 μm along its direction of travel. At $35°$ incidence angle the laser will have covered/moved laterally along the sample surface 5.35 μm. Comparing the entire area probed by the electron pulse (4.68×10^4 μm^2) and the area irradiated/excited by the laser pulse (256 μm^2) the ratio of excited to non-excited sample area is 0.005. The precise analysis would be to consider how much volume was probed/excited, but the ratio will be almost identical. Thus, the information the electron pulse will carry onto the detector will primarily be information pertaining to the undisturbed sample, 99.5 %, and only 0.5 % of the electron signal will be containing information about the reaction that is just starting to unfold in the sample on one edge. The observer will, in essence, see no changes in the sample's diffraction pattern or image. It is important to remember that once the laser photons interact with the sample there is a sequence of events that take place (see section Ultrafast Conditions in Matter) and that it usually takes about 100 fs for the excited electron gas to thermalize before anything else can start happening in the sample (at least on the nanoscale—in a 50 nm thick sample). Consequently, the laser energy that was transferred to the sample in those first 10.26 fs will not have caused any significant changes to the sample.

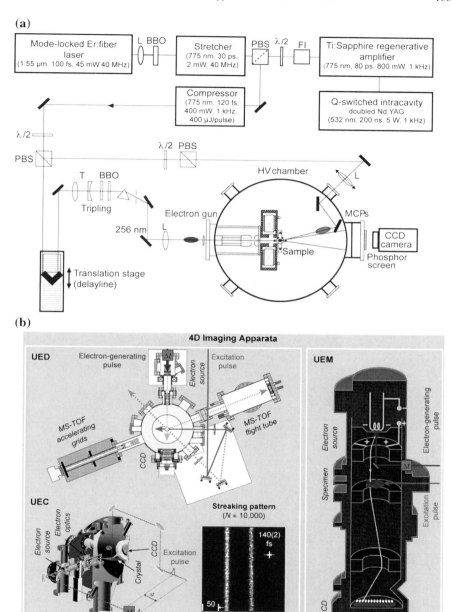

Fig. 4.9 Ultrafast electron imaging and diffraction apparata at University of Montreal, California Institute of Technology, and Lawrence Livermore National Laboratory: **a** A system configured for ultrafast transmission electron diffraction (UED) (courtesy Dr. B. J. Siwick [57]), **b** Ultrafast electron diffraction (UED) and ultrafast electron crystallography (UEC) and ultrafast electron microscopy (UEM). Shown also are the electron streaks (inset) for10000 electrons (courtesy Dr. A. H. Zewail [58]), **c** Ultrafast transmission electron microscope (courtesy Dr. G. H. Campbell [57])

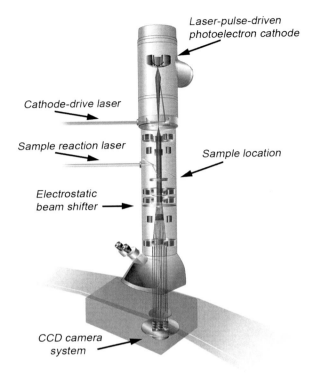

Fig. 4.9   continued

Let's move more than 100 fs ahead to $t_1 = 233.5$ fs elapsed time, to make sure a detectable reaction is ongoing in the sample. The laser pulse front will have moved now 122.07 μm laterally across the sample and it propagated 70.02 μm in its direction of travel. The laser has been depositing energy for 233.5 femtoseconds, but unfortunately not simultaneously in the entire illuminated area, but rather consecutively as it swept across the sample surface. The first point of contact between the laser pulse and the sample has advanced the most in terms of reaction progress, whereas the position where the laser is now, at t_1, just received its first photons depositing energy. If the electron pulse were to enter the sample surface now and again spend 10.02 fs to traverse the entire sample thickness collecting information, it will encounter the situation of an ongoing reaction in one half of the sample, whereas the other half that has not been irradiated by the laser yet, is still in its original state. The ratio of excited to non-excited sample area is now much better, 50 % instead of 0.5 % as in the previous case. However, the excited sample volume reflects a reaction development that has been trailing the laser pulse front sweeping across the sample surface. All reaction stages that will develop within those 233.5 fs will now be present in the excited half of the sample, but the electron pulse will collect, in one single snapshot, all the information within its 10.26 fs long presence in the sample and transfer it all in one, single diffraction pattern or image onto the detector. In essence, it will be smearing all the relevant information of the individual reaction stages into a single detection signal.

Moving the laser pulse front to $t_2 = 467$ fs—the end point of where the laser is depositing its last quantum of energy into the sample—an electron pulse entering the sample surface now will just collect, in an identical manner as before, information that pertains to the continuing reaction in the sample, now ongoing for 467 fs. As a result the time resolution that one hopes to attain is, by far, not defined anymore by the electron or laser pulse duration, instead it is now defined by geometrical factors. Determining the moment of the true "time zero" is thus difficult, as there will be always one certain area of the sample that has just received its share of laser energy and is at the beginning of the reaction to be triggered.

There are a few ways to remedy this difficult scenario. One obvious is to reduce the relative angle between laser and electron path to the minimum possible (ideally zero degrees) to achieve collinearity. However, this is not always possible in the real experimental setup. Another way is to reduce the electron pulse's lateral dimensions to just probe the small area that is being excited by the laser front. In the presented scenario above, the laser really only illuminates a thin line (244.1×5.2 μm in dimension) of the sample as it comes in at an angle and sweeps across. The electron pulse would therefore need to be focused down to 5.2 μm to just probe the currently excited sample volume. It could certainly be delayed in time to probe the same sample area after the laser has deposited its energy, that is, later stages of the reaction. However, it will be quite challenging to focus the electron pulse containing at least 10^8 electrons to these dimensions and still have no Coulomb repulsion active. This may well destroy all focusing and consequently all probing efforts. Making instead the laser pulse's lateral dimensions larger, which is physically very easy, will not be helpful, as one still has the 5.2 micrometer thin line illuminating and exciting the sample volume. Switching beam paths, that is, the electron pulse is incident at an angle and the laser pulse comes in perpendicular, certainly ensures that the laser energy is deposited simultaneously over the entire irradiated area and there is no reaction trailing behind the laser front, but the electron pulse will have only a small fraction of its electron population probing each consecutive stage of the reaction. As the electron pulse traverses the sample the reaction is unfolding continuously. In this arrangement it would take the electron pulse 746.8 fs to cross the entire sample and it would capture information belonging to every stage of the reaction unfolding for that amount of time—a very long time compared to the original electron pulse duration of only 10 fs.

The current state-of-the-art instrumentation does not quite permit forming 10 fs short electron pulses hence this problematic situation described does not really occur yet, however, it will not take long until 10 fs short electron pulses can be generated. Current experiments are being performed with 1 ps electron pulses and 10–100 fs short laser pulses. A 1 ps long electron pulse, accelerated to 100 kV, incident perpendicular onto the sample, will physically be 187.5 μm thick and therefore need a bit more than 1 ps to traverse the entire sample, front to back. As above, the 10 fs short laser pulse will have traversed the entire sample in 467 fs, more than twice as fast as the electron pulse. It will have deposited all of its energy, excited the reaction and this one would already be ongoing for 1 ps until the electron pulse exits the sample with its last bit of information pertaining to a reaction that has been ongoing for at least 533 fs (since the last energy deposition by the laser). Again, the information that the electron pulse is carrying to the detector is the entire collection of all reaction stages; a composite of information. It makes the experimental results blurred in the sense that information, structural information for example, of all reaction stages is printed into one and the same diffraction pattern or image. However, there are experimental results that show a fairly strong and clear diffraction pattern of such ongoing reactions, not necessarily a composite of all the possible information. This may well be due to non-linear reaction trajectories and intermittent stages that can only develop once certain conditions are met within the sample. Remember, that approximately within the first 100 fs, for example, the excited electron gas needs to thermalize first before it can transfer its energy to the surrounding crystal structure, such that a physically measurable change can start taking place in the sample. The progress of a reaction is therefore not linear in time (in macroscopic terms: it may stall for a brief moment in one intermittent stage), and as the electron pulse traverses the sample—fairly linear in time (absorption and multiple scattering neglected)—it may pick up information of one particular reaction stage, more than of any other. Hence, the resulting diffraction pattern, for example, can exhibit clear signs of this particular intermittent reaction stage, but it never presents solely this stage. It is always blurred by information belonging to a previous or a following reaction stage.

The last viable way to remedy the discussed situation is to use a tilted laser pulse front impinging on the sample surface. A similar approach has been proposed for use in ultrafast electron crystallography (UEC). There the group velocity mismatch between electron and laser pulse is the critical parameter [56]. For UEM and UED a tilted laser pulse front may come in at almost any desired angle of incidence, as long as the laser pulse front can be tilted to counterbalance that angle, (see Fig. 4.10).

The idea is to have both the laser pulse and the electron pulse impinge on the sample surface in a flat manner. This condition has the advantage of having the laser excite the entire sample area to be probed by the electron pulse simultaneously, as if it were to come in collinear with the electron pulse. The sample would not show any trailing of a reaction across its laser-irradiated area. Any time delay between the laser and the electron pulse would now enable to solely probe the reaction stage currently ongoing in the sample, whether it is an intermittent (stalled) or a highly transient stage. The resulting diffraction patterns or images would contain only that information and not be a composite of information anymore.

The physical realization of such tilted laser pulse front is in principle an easy task, however, there are a few drawbacks to consider. Any laser pulse, as monochromatic as it may be, exhibits a spectrum of wavelengths—the shorter the pulse the broader the spectrum. There are two methods to perform a laser pulse front tilt: one is using a grating, whereas the second makes use of a prism. Both methods will create a tilted laser pulse front that exhibits angular dispersion and spatial

Fig. 4.10 Identical
experimental setup as in
Fig 4.8, but with a tilted
laser pulse impinging on
the sample surface in a flat
manner, coincident with the
electron pulse

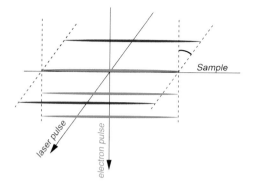

chirp. The latter one is a frequency variation transversely across the pulse (long wavelengths on one end, short wavelengths on the opposite end). Angular dispersion, however, is not particularly desired for ultrafast experiments with critically defined parameters. Especially the broadband character of these ultrashort laser pulses makes them particularly vulnerable to angular dispersion, as it tends to erode the temporal resolution, degrade pulse shape, and thus reduce intensity. Therefore, intentionally introducing angular dispersion to the laser pulses is to be regarded with caution. In fact, it generally is often an irritating phenomenon when working with ultrafast lasers. On the other hand angular dispersion serves our purpose to obtain a tilted laser front. Note though, that prisms will generally disperse light over a much larger frequency bandwidth than diffraction gratings, making them useful for broad-spectrum spectroscopy. Additionally, prisms do not suffer from complications arising from overlapping spectral orders, which all gratings have. In a prism, due to the group-velocity dispersion in the medium, the redder end (longer wavelength) will be emerging earlier than the bluer end (shorter wavelength). Hence, one has created a tilted pulse front—with spatial chirp and angular dispersion. The latter one needs to be reversed though. This can be done, for example, by a spherical mirror that focuses the broadening pulse back to its original diameter and simultaneously maintains the tilted pulse front. This setup requires very careful and precise alignment of all optical components to precisely match the laser pulse tilt angle to the relative angle between the laser and the electron pulse. Slight misalignments will only cause loss in temporal and instrumental resolution. It is important to note that the angular dispersion was introduced by the prism, and it was only reversed, in fact, inversed but not eliminated by the spherical mirror. Once the focal point/crossover is passed the pulse will be dispersing again. Thus, the further the tilted pulse travels the more separation of the frequency components will occur and the larger the pulse dimension (transverse). Clearly, a pulse shape changing mechanism that is not desired for ultrafast experiments.

To truly eliminate angular dispersion, one can use a second prism and align it anti-parallel to the first one. Since the magnitude of the group velocity dispersion

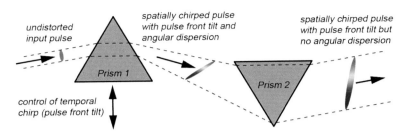

Fig. 4.11 A two-prism anti-parallel configuration that allows to conveniently adjust the pulse front tilt

depends on the propagation distance in the dispersive medium—in the first prism, the shorter wavelengths (blue) have the longer paths, whereas in the second prism the longer wavelengths (red) have the longer path to travel through the dispersive medium—this configuration will effectively eliminate the angular dispersion introduced by the first prism. Unfortunately, it does not remove the spatial chirp. But more importantly, it allows controlling the magnitude of the pulse front tilt by intentionally adjusting the relative position of the anti-parallel prisms (Fig. 4.11).

Pulse front tilt and angular dispersion are generally though to be equivalent phenomena [59–61], however, it has been shown that pulse front tilt can be achieved without angular dispersion, just with a combination of spatial and temporal chirp [62]. Spatial chirp, as explained above, is characterized by the frequency gradient $\varphi = d\omega_0/dx$, where ω_0 is the mean frequency at position x, whereas temporal chirp is characterized by the group velocity dispersion. For the described setup to produce a pulse front tilt without angular dispersion it requires an incoming pulse with spatial chirp (on the second prism). Adding temporal chirp will then allow to conveniently adjust the magnitude of the pulse front tilt and it therefore facilitates a precise matching of the pulse front tilt to the relative angle between the electron and the laser pulse.

The only drawback to this matching-of-angles-via-laser-pulse-front-tilt method is the remaining spatial chirp in the laser pulse impinging on the sample surface. Remember that spatial chirp is a frequency variation transversely across the pulse (long wavelengths on one end, short wavelengths on the opposite end). The consequence of this is an energy gradient transversely across the pulse, and the laser pulse will therefore deposit more energy on one end of the irradiated area and less on the opposite end. The reaction to be triggered and observed in the sample might require a (minimum) threshold energy to start developing, however, this may vary or not be as critical depending on the type of sample and what kind of reaction is to be investigated. If a minimum energy is required to get a reaction going, as for example in photoemission and the associated work function, one could adjust the laser minimum energy to exceed this threshold and without doubt disregard the surplus energy provided on the other side of the laser pulse.

4.6 Experimental Variations and Ultrafast Electron Detection

4.6.1 Single- Versus Multi-shot Experiments

In single-shot experiments one has one single chance to capture all the relevant information pertaining to an ultrafast reaction. The laser pulse deposits its energy once, unfolds the reaction, and the camera has to be quick enough to take snap-shots during the process—ideally with shorter exposure times than the duration of the entire reaction path. The reaction that is of interest is usually irreversible and the sample may also be destroyed afterwards. Therefore, one has no second chance to repeat the experiment on the same sample spot, for example, on a thin foil. The only option for a repetition is to move to a new area on the sample and perform the same experiment again. Sample variations as composition, thickness, and location of the irradiated area have to be taken into account when compar-ing the results of multiple such single-shot experiments. However, instead of using a one-laser-pulse—one-electron-pulse configuration of the experiment one may opt for a one-laser-pulse—three-electron-pulse arrangement. The laser that helps generating electron pulses at the cathode can be pulsed in such manner that for example three short consecutive laser pulses hit the cathode at adjustable inter-vals, thus creating a train of electron pulses, each for example, 1 ps in duration and 10 ps apart. The feasibility certainly depends on the time frame of the reaction. The reaction needs to be ongoing for longer than the time span of the train of elec-tron pulses. However, there are many ultrafast reactions where this arrangement works out well, yielding results that give insight into the dynamics of the reac-tion. In essence, one has then a "three frame movie" after the experiment has com-pleted. The irreversible nature of such reactions demand that all the electron pulses are loaded with as many electrons as can be generated, compressed, manipulated, etc. to have maximum electron-sample interaction, collection of information and a decent signal-to-noise ratio. Basically, a multi-electron single-shot experiment. The problems associated with all these boundary conditions were discussed in pre-vious sections and paragraphs of this chapter.

Multi-shot experiments on the other hand have the freedom to not be bound by the irreversible nature of the reaction. The reactions are reversible and as men-tioned before single-electron pulses can be used. The rather cyclic experiment is a single-electron multi-shot arrangement. The repetitive reaction in the sample can be examined only because every time the reaction is initiated a single-electron scattering (probing) event is recorded, contributing each time to the completion of the final picture, either in diffraction or in imaging mode.

Both types of experiments require similar, yet different electron detectors. A more detailed description of the types of currently available electron detectors for ultrafast experiments is provided in Chap. 7 of this book. This chapter instead, will describe the techniques how such detectors can be used in UED and UEM.

4.6.2 *Electron Detection Techniques*

Obviously, the parameters that are critical to UED and UEM experiments are speed, sensitivity, and the dynamic range of the detector. To date there is no direct electron detector that can match the speed of the laser or the electron pulses. The readout speed of the camera depends on internal electronics, how large the imaging area is, whether or not an integrated frame store area is part of the detector design, readout and reset mode, and a few other variables. Under optimum conditions state-of-the-art detectors that are suitable for such experiments can reach tens of microsecond time resolution, that is, the readout time per frame. However, employing suitable techniques can yield picosecond time resolution.

One established technique to detect extremely fast electron pulse snapshots in time is to revert to diffraction patterns and record them on steak cameras. As described in a previous section of this chapter, streak cameras operate by employing a time-varying deflection of the incident electron pulse—in this case the diffraction pattern. A quickly modulated electric field between a pair of deflection plates serves this purpose and the diffraction pattern is swept over a linear detector, usually a charge-coupled device (CCD). During an ultrafast process the spots in the associated diffraction pattern may either change relative distance, shape or intensity. All these alterations are, in principle, detectable with such a streak camera. In reality however, it is experimentally very difficult and the outcome is especially limited by the dynamic range of the linear detector. Finally, the interpretation of the results is very challenging. The time dependent changes of position, intensity and form (width) of the diffraction spots reveals information about structural evolutions. Increased atomic motion (or separation) due to ultrafast heating causes a shift of the diffraction spots in reciprocal space towards lower values, concomitant with an increase in the diffraction spot width, and a reduction of spot intensity. Spot displacement can in principle be monitored with such a streak camera. Intensity fluctuation instead, can be a bit more challenging, and rely strongly on the performance, especially the sensitivity and dynamic range of the detector. A fundamental concept in crystallography is the structure factor. A kinematical approximation for diffraction gives for the intensity I of a diffracted beam.

$$I_{\Delta k} = |\psi_{\Delta k}|^2 \propto |F_{\Delta k}|^2 \tag{4.4}$$

$$F_{\Delta k} = \sum_j f_j e^{-i\Delta k \cdot r_j} \tag{4.5}$$

where $\psi_{\Delta k}$ is the wavefunction of the beam scattered a vector Δk, $F_{\Delta k}$ represents the structure factor, r_j is the position $(xyz)_j$ of atom j in the unit cell, and f_j is the scattering power of that particular atom. Observation of the intensity of a specific diffraction spot will therefore reveal the unique motions of atoms drawn into the dynamic process.

Another technique that has proven useful for capturing UEM images (it is also appropriate for recording diffraction patterns), applies a deflection field to the (three) individual consecutive electron pulses used in a multi-electron single-shot configuration. In essence, this allows splitting the full camera frame into smaller sections (half, thirds, quarters) by shifting each electron pulse to a corresponding area on the camera frame that has not been exposed yet. Drawbacks to this technique, is the reaction time of the electro-magnetic and deflection plates lenses that redirect the electron pulse(s). They tend to be in the lower nanosecond range and thus making this technique not suitable to record faster reactions and processes. Furthermore, it is not quite established how much trailing of the electron pulse to the changing deflection potential is involved in this technique.

4.7 Conclusion

The developments in ultrashort pulsed electron based methods for the pump-probe as well as for related ultrafast types of experiments have reached a point, where pico- and even femtosecond short electron pulses are attainable and controllable to a certain degree to monitor and follow atomic motions during an ultrafast process. The instrumental modifications that need to be accomplished in terms of electron gun design, electron emitter material, pulse propagation, space charge effects in electron pulses, pulse compression and characterization techniques, synchronization and experimental geometric considerations have been presented and discussed. The field of UEM and UED has by far not reached the status of an established characterization technique and multiple innovations and developments are still required to make this a viable and reliable method to assess the dynamics of ultrafast reactions that are moving more and more into the focal point. Scientists have come to realize that static high resolution, and even three-dimensional structural information is not sufficient anymore to satisfy the need to understand structure-property and especially, structure–function relationships.

References

1. W. S. Fann, R. Storz, H.W.K. Tom, J. Bokor, Phys. Rev. B 46, 13–592 (1992)
2. D.M. Riffe, X.Y. Wang, M.C. Downer, D.L. Fisher, T. Tjima, J.L. Erskine, R.M. More, J. Opt. Soc. Am. B **10**, 1424 (1993)
3. M. Merschdorf, W. Pfeiffer, S. Voll, G. Gerber, Phys. Rev. B **68**, 155416 (2003)
4. Y.C. Martin, H.F. Hamann, H.K. Wickramasinghe, J. Appl. Phys. **89**, 5774 (2001)
5. A.W. Baum, J.E. Schneider, R.F.W. Pease, M.A. McCord, W.E. Spicer, K.A. Costello, V.W. Aebi, J. Vac. Sci. Tech. B **15**, 2702 (1997)
6. European Patent No.EP0834896B1, filed 02.10.1996, granted 12.06.2002 to Rainer Spehr, Michael Schmitt, Jürgen Frosien
7. B.J. Siwick, A.A. Green, C.T. Hebeisen, R.J.D. Miller, Opt. Lett. **30**, 1057 (2005)
8. B.J. Siwick, J.R. Dwyer, R.E. Jordan, R.J.D. Miller, Chem. Phys. **299**, 285 (2004)
9. B.J. Siwick, J.R. Dwyer, R.E. Jordan, R.J.D. Miller, J. Appl. Phys. **92**, 1643 (2002)
10. T. van Oudheusden, E.F. de Jong, S.B. van der Geer, W.P.E.M. Op't Root, O.J. Luiten, B.J. Siwick, J. Appl. Phys. **102**, 093501 (2007)
11. P. Baum, A.H. Zewail, Chem. Phys. Lett. **462**, 14 (2008)

12. A.H. Zewail, Annu. Rev. Phys. Chem. **57**, 65 (2006)
13. P. Baum, D.S. Yang, A.H. Zewail, Science **318**, 788 (2007)
14. R. Srinivasan, V.A. Lobastov, C.Y. Ruan, A.H. Zewail, Helv. Chim. Acta **86**, 1763 (2003)
15. H. Dömer, O. Bostanjoglo, Rev. Sci. Instrum. **74**, 4369 (2003)
16. H. Dömer, O. Bostanjoglo, Microsc. Microanal. **9**, 358 (2003)
17. H. Dömer, O. Bostanjoglo, Adv. Eng. Mater. **4**, 623 (2002)
18. H. Dömer, O. Bostanjoglo, Appl. Surf. Sci. **208**, 442 (2003)
19. T. LaGrange, G.H. Campbell, P.E.A. Turchi, W.E. King, Acta Mater. **55**, 5211 (2007)
20. T. LaGrange, Appl. Phys. Lett. **89**, 044105 (2006)
21. G.H. Campbell, Microsc. Microanal. 12, 1428CD (2006)
22. N.D. Browning, Microsc. Microanal. 12, 790CD (2006)
23. T. LaGrange, G.H. Campbell, J.D. Colvin, B. Reed, W.E. King, J. Mater. Sci. **41**, 4440 (2006)
24. M.R. Armstrong, K. Boyden, N.D. Browning, G.H. Campbell, J.D. Colvin, W.J. DeHope, A.M. Frank, D.J. Gibson, F. Hartemann, J.S. Kim, W.E. King, T.B. LaGrange, B.J. Pyke, B.W. Reed, R.M. Shuttlesworth, B.C. Stuart, B.R. Torralva, Ultramicroscopy **107**, 356 (2007)
25. H. Kleinschmidt, A. Ziegler, G.H. Campbell, J.D. Colvin, O. Bostanjoglo, J. Appl. Phys. **98**, 054313 (2005)
26. O. Bostanjoglo, Adv. Imaging Electron Phys. **121**, 1 (2002)
27. S.I. Anisimov, B.L. Kapeliovich, T.L. Perelman, Sov. Phys. JETP **39**, 375 (1974)
28. M.I. Kaganov, I.M. Lifshitz, L.V. Tanatarov, Sov. Phys. JETP **4**, 173 (1957)
29. B. Rethfeld, A. Kaiser, M. Vicanek, G. Simon, Proc. SPIE **3343**, 388 (1998)
30. B. Rethfeld, H. Krutsch, D.H.H. Hoffmann, Contrib. Plasma Phys. **50**, 16 (2010)
31. T. Feuerer, H. Langhoff, Appl. Phys. A **63**, 13 (1996)
32. S. Nolte, C. Momma, B.N. Chichkov, H. Wellig, Phys. Blätter **55**, 41 (1999)
33. G.A. Mourou, T. Tajima, S.V. Bulanov, Rev. Mod. Phys. **78**, 309 (2006)
34. T.H. Her, R.J. Finlay, C. Wu, S. Deliwala, E. Mazur, Appl. Phys. Lett. **73**, 1673 (1998)
35. T.H. Her, R.J. Finlay, C. Wu, E. Mazur, Appl. Phys. A **70**, 383 (2000)
36. C. Wu, C.H. Crouch, L. Zhao, J.E. Carey, R. Younkin, J.A. Levinson, E. Mazur, R.M. Farrell, P. Gothoskar, A. Karger, Appl. Phys. Lett. **78**, 1850 (2001)
37. F. Sanchez, J.L. Morenza, R. Aguiar, J.C. Delgado, M. Varela, Appl. Phys. Lett. **69**, 620 (1996)
38. A.J. Pedraza, J.D. Fowlkes, D.H. Lowndes, Appl. Phys. Lett. **77**, 3018 (2000)
39. A.J. Pedraza, J.D. Fowlkes, S. Jesse, C. Mao, H. Lowndes, Appl. Surf. Sci. **168**, 251 (2000)
40. A.J. Pedraza, J.D. Fowlkes, D.H. Lowndes, Appl. Phys. Lett. **74**, 2322 (1999)
41. R.J. Younkin, J.E. Carey, E. Mazur, J.A. Levinson, C.M. Friend, J. Appl. Phys. **93**, 2626 (2003)
42. B. Koopmans, G. Malinowski, F.D. Longa, D. Steiauf, M. Fähnle, T. Roth, M. Cinchetti, M. Aeschlimann, Nat. Mater. **9**, 259 (2010)
43. D. Ashkenasi, A. Rosenfeld, H. Varel, M. Waehmer, E.E.B. Campbell, Appl. Surf. Sci. **120**, 65 (1997)
44. B. Luk'Yanchuk, N. Bityurin, S. Anisimov, D. Baeuerle, Appl. Phys. A 57, 367 (1993)
45. P.T.C. So, H. Kim, I.E. Kochevar, Opt. Exp. **3**, 339 (1998)
46. N. Shen, D. Datta, C.B. Schaffer, P. LeDuc, D.E. Ingber, E. Mazur, Mech. Chem. Biosyst. **2**, 17 (2005)
47. A. Heisterkamp, I.Z. Maxwell, E. Mazur, J.M. Underwood, J.A. Nickerson, S. Kumar, D.E. Ingber, Opt. Exp. **13**, 3690 (2005)
48. K. Koenig, I. Riemann, P. Fischer, K. Halbhuber, Cell. Mol. Biol. **45**, 192 (1995)
49. U.K. Tirlapur, K. Koenig, Nature **448**, 290 (2002)
50. W. Watanabe, N. Arakawa, S. Matsunaga, T. Higashi, K. Fukui, K. Isobe, K. Itoh, Opt. Exp. **12**, 4203 (2004)
51. A. Vogel, V. Venugopalan, Chem. Rev. **103**, 577 (2003)
52. M. Martyniuk, J. Antoszewski, C.A. Musca, J.M. Dell, L. Faraone, Smart Mater. Struct. **15**, S29 (2006)

53. J. Cao, Z. Hao, H. Park, C. Tao, D. Kau, L. Blaszczyk, Appl. Phys. Lett. **83**, 1044 (2003)
54. Z. Chang, A. Rundquist, J. Zhou, M.M. Murnane, H.C. Kapteyn, X. Liu, B. Shan, J. Liu, L. Niu, M. Gong, X. Zhang, Appl. Phys. Lett. **69**, 133 (1996)
55. J.C. Diels, W. Rudolph: Ultrashort Laser Pulse Phenomena: Fundamentals, Techniques, and Applications on a Femtosecond Time Scale, 2nd edn. (Academic Press, Elsevier, 2006)
56. P. Baum, A.H. Zewail, Proc. Nat. Acad. Sci. **103**, 16105 (2006)
57. W.E. King, G.H. Campbell, A. Frank, B. Reed, J.F. Schmerge, B.J. Siwick, B.C. Stuart, P.M. Weber, J. Appl. Phys. **97**, 111101 (2005)
58. D. Shorokhov, A.H. Zewail, Phys. Chem. Chem. Phys. **10**, 2879 (2008)
59. Z. Bor, B. Racz, G. Szabo, M. Hilbert, H.A. Hazim, Opt. Eng. **32**, 2501 (1993)
60. J. Hebling, Opt. Quantum Electron **28**, 1759 (1996)
61. C. Dorrer, E. M. Kosik, I. A. Walmsley, Appl. Phys. B 74(suppl.), 209 (2002)
62. Z. Bor, B. Racz, Opt. Commun. **54**, 165 (1985)

Chapter 5
In-situ and Kinetic Studies Using Neutrons

Götz Eckold and Helmut Schober

Abstract Neutrons provide unique possibilities for in-situ studies of condensed matter due to their ability to penetrate large samples and work pieces, to distinguish between neighboring elements and even between isotopes of the same element, to interact with magnetic moments and to map nuclear and magnetic excitations. The range of applications extends from biological or soft matter systems over basic superconducting and magnetic investigations to all kinds of materials science under a great variety of external conditions. Complex and sophisticated sample environments have been developed in the past to study samples under external perturbations and extreme conditions like high or low temperatures, high pressures and mechanical stresses, electric and magnetic fields and different kinds of chemical environments. Not only equilibrium studies are in the focus of neutron scattering investigations—increasing interest in real-time kinetic studies lead to new techniques that allow even inelastic studies on time scales down to the microsecond regime thus providing most direct information about the changing chemical bonds in materials. In the present chapter, we want to provide an overview over recent developments for in-situ studies using neutrons and to highlight some application in materials science without having the ambition to give a comprehensive review. Selected examples of neutron studies under extreme conditions from a variety of different disciplines are collected to demonstrate the power of neutron scattering for materials characterization. Moreover, kinetic studies providing information about the relaxation behavior of materials under external load are presented covering time-scales

G. Eckold (✉)
Institut für Physikalische Chemie, Georg-August-Universität Göttingen, Tammannstrasse 6, 37077 Göttingen, Germany
e-mail: geckold@gwdg.de

H. Schober
Science division, Institut Laue-Langevin, 6, rue Jules Horowitz, BP 156, 38042 Grenoble Cedex 9, France
e-mail: schober@ill.eu

A. Ziegler et al. (eds.), *In-situ Materials Characterization*,
Springer Series in Materials Science 193, DOI: 10.1007/978-3-642-45152-2_5,
© Springer-Verlag Berlin Heidelberg 2014

from hours or days down to the microsecond regime. Those investigations are the basis for the understanding of the basic microscopic mechanisms and allow one to tailor materials using non-equilibrium states.

5.1 Introduction

Neutron scattering has become a standard tool for investigating condensed matter. Like any scattering technique it extracts information about the structure and dynamics of the sample from the momentum and energy distribution of the scattered particle. The neutron interacts with the nuclei and magnetic moments present in a material. This interaction has exactly the right strength for the investigation of bulk samples. As a rule of thumb, we aim for a scattering probability in the 10 % range. This value assures that we have on one hand a sufficiently intense scattered signal and that on the other hand only a small fraction of that signal originates from multiple scattering. Multiple scattering is detrimental to the data analysis as it destroys the correspondence of momentum (Q) and energy (E) change connected with a single scattering event on one side and the detectable overall energy and momentum change on the other side. For a thermal neutron beam 10 % of scattering corresponds in the case of Aluminium to passing through one billion layers of atoms or about 1 cm of sample. In the case of hydrogenated samples this thickness has to be reduced by about a factor 100 but remains a fraction of a mm. The weak scattering power combined with generally weak absorption leads to a high penetration power of neutrons. This allows for a straightforward use of even complex and bulky sample environment. The low absorption has the additional benefit of depositing little energy per unit volume of sample. Thermal neutrons contrary to intense X-ray beams thus constitute a delicate probe of matter that is adapted even to fragile materials.

The energies of neutron beams that are available for materials research range from a few neVs up to eVs. The corresponding neutron wavelengths cover the full range from the sub-atomic (fractions of an Å) to the super-molecular spacing (several nm). Thus while the wavelengths are comparable to those of X-rays the energies are orders of magnitude smaller. This is due to the fact that the neutrons have been thermalized in the moderators of the neutron sources and hence possess energies corresponding to the moderator temperatures. As a consequence, neutron energies match the typical energies of excitations at these temperatures. This facilitates dynamic studies, as the energy exchanged with the sample is generally a good fraction of the initial energy of the neutron. Neutrons thus provide access to the slow motion of polymers as well as the fast excitations in the electronic spin system.

A particularity of neutrons is the fact that they interact with the nuclei and are, therefore, sensitive to isotopes. As isotopes have little influence on the chemistry this sensitivity can be exploited for contrast variation. The most prominent example is the total or partial substitution of hydrogen by deuterium. This so-called deuteration is at the very heart of many investigations in the fields of

Table 5.1 Characteristic performance parameters of neutrons

Size and sensitivity	Time scales
Smallest sample volume 0.001 mm^3	Shortest time detectable 10^{-15} s
Sensitivity to displacements 10^{-4} nm	Longest relaxation times meas. 10^{-6} s
Chemical sensitivity 10 ppm	Time resolution for kinetics $<10^{-3}$ s
Magnetic sensitivity 0.01 μB/atom	Stability in time: years

soft condensed matter and biology. The coupling of the spin of the neutron to microscopic magnetic fields makes the neutron a unique probe to decrypt even the most complex magnetic structures and fluctuations.

There has lately been a strong boost in performance mainly due to instrument upgrades enabled by new technology. These upgrades lead to an increased sensitivity of the spectrometers. It is now possible to investigate samples as small as 0.001 mm^3 (for other performance parameters see Table 5.1). The increased flux is particularly useful when it comes to kinetic studies that are the subject of this article.

Modern materials investigations require sophisticated sample environment. It is important that the microscopic picture given by the neutron scattering experiment reflects the properties of the sample under well-defined conditions. While the in-situ study of systems under extreme conditions like temperatures in the nano-Kelvin regime or above 2000 K, pressures of several 100 GPa, magnetic fields in the 20 T-regime or electric fields of several 10 kV/cm provide new information about the properties of materials in thermodynamic equilibrium, time-resolved investigations allow monitoring the evolution of a system towards equilibrium. If a system is exposed to a sufficiently strong external perturbation, the reaction path can be tracked by microscopic methods like diffraction or scattering. Quite frequently, intermediate states are detected that are meta-stable in character but exhibit unusual and exciting properties. If required these intermediate states can be stabilized over large time-intervals by appropriate (thermal) treatments like quenching thus leading to new non-equilibrium materials.

In the following, we want to illustrate the possibilities of neutron techniques for the characterisation of transition states and kinetic behaviour. It is not the aim of this article to provide a comprehensive review of kinetic studies with neutrons. Rather, we want to highlight recent developments in this field using some selected examples.

There are two fundamentally different kinds of kinetic studies: Single shot experiments, on the one hand, are usually restricted to processes or reactions that occur on time-scales adapted to the counting times of neutron experiments, i.e. seconds to hours. With stroboscopic experiments, on the other hand, one can access much shorter time scales down to the microsecond regime at least, but these studies are restricted to repeatable processes and can also be considered as pump-and-probe experiments.

The unique property of the neutron as a strongly penetrating probe allows using sophisticated sample environments to produce the required perturbations of the system from equilibrium. More precisely, extreme conditions of temperature, pressure or stress, electric and magnetic fields as well as chemical potential can be realized—and this not only statically, but also in a cyclic way that allows for stroboscopic measurements.

5.2 Conditions Available for In-situ Studies

The main aim of this article is to give insight into the possibilities of investigation that neutrons offer for kinetic studies. Before doing so we would like to review briefly the conditions available for in-situ studies in general. Whether these conditions can be applied to kinetic studies will then depend as outlined above on the counting-times and possibilities of cycling.

5.2.1 Temperature

Using standard techniques, temperatures between a few mK and 1500 K are easily accessible at almost every type of neutron instrument, from diffractometers over small angle instruments to spectrometers. There are, however, dedicated developments for experiments of lowest temperatures down to the nanokelvin range, which is particularly interesting for the investigation of small magnetic interactions leading e.g. to the ordering of nuclear spins. One of the main activities in this lowest temperature science with neutrons is concentrated at the BER-II research reactor in Berlin [1].

On the other extreme, new devices are being developed that allow performing neutron scattering experiments on large samples at temperatures well above 2500 K. Particularly promising are acoustic or electromagnetic levitation techniques that allow for measurements without containers [2–4].

Subjects of investigation are the structure and dynamics of liquids and melts as well as anharmonic effects or disorder phenomena in high-temperature materials.

5.2.2 Pressure and Stress

Hydrostatic pressures up to 1 GPa are routinely available on nearly all instruments. This value is small compared to the pressures available with diamond-anvil cells at X-ray facilities. The difference is explained by the larger samples needed for neutron experiments. The so-called Paris-Edinburgh-cell was developed to meet the requirement of large volume while considerably extending the pressure range. Pressures up to 10 GPa at sample volumes of 100 mm^3 and temperatures from 80 to 1000 K can thus be realised in routine operation [5–8]. The application of hydrostatic pressure is of particular interest for geological applications. Moreover, the origin of negative thermal expansion of materials can also be studied by high-pressure inelastic work since the variation of phonon frequencies with pressure determines the Grüneisen-parameters.

High-pressure work has been identified as a very important and growing field of neutron science. In the US, the Los Alamos Neutron Science Centre LANSCE has made great efforts to provide specialised devices for the use with different neutron instruments [9]. At the FRM II in Garching, a multi-anvil press is being installed at the new time-of-flight diffractometer. This huge device will enable experiments und geologically relevant conditions, i.e. pressures up to 25 GPa and temperatures up to 2800 K for sample volumes of 20 mm^3. Moreover, The multi-anvil design allows one to perform dedicated deformation experiments since all anvils can be individually driven.

Another kind of in-situ experiments under mechanical load at stationary conditions are rheological studies on soft matter with small angle neutron scattering. Stationary shear stresses can be applied by rotating cups in the Couette configuration or by using a plate–plate geometry, e.g., in order to investigate the structural changes of polymers, micellar solutions or gels. A dedicated Neutron Rheometer for high torques up to 200 Nm or high shear rates up to 865 s^{-1} has been commissioned at the Los Alamos Science Centre recently [10]. For high-viscosity systems a rheometer using rotating plates was installed at the ISIS Spallation Source at Rutherford Appleton Laboratory [11].

Externally applied stress is equally a very important parameter in classical materials engineering investigations. During the past decade, diffractometers dedicated to the determination of stress from the measurement of lattice strain have been developed which can host complete mechanical components or structures. They enable in-situ studies for the determination of materials properties, manufacturing parameters or in service behavior, simulating real operational conditions. By measuring the strain response as a function of applied stress the (diffraction) elastic constants of a material can be determined. These parameters are vital for the calculation of stresses from lattice strain measurements. In the same manner, in-situ deformation studies of materials can be performed. Since diffraction is crystal orientation and phase selective, the study of deformation mechanisms and load sharing in single phase, multiphase or composite materials is possible. Other important investigations performed are on crack development, creep or fatigue behavior of a material by applying constant or cyclic loading during a diffraction experiment [12–14].

Additionally, investigating phase fractions, stress and texture during processes such as welding is a challenging experimental work, since an automated welding bench has to be operated on the diffractometer. However initial in-situ studies have already been performed on fusion and friction welding during neutron diffraction experiments [15].

5.2.3 Magnetic Fields

Cryomagnets with split coil geometry are frequently used for neutron scattering experiments. The combination of strong fields and very low temperatures allows the study of magnetic ordering phenomena and details of complex magnetic interactions. The simultaneous use of polarized neutrons provides even more information. Nowadays, magnets with fields up to 7–10 T belong to the standard equipment of modern neutron facilities and even 15 T-split coil magnets are available.

To go beyond these field strengths requires extremely great efforts since the design of those magnets always has to provide enough space for incoming and scattered neutrons, thus leading to enormous forces between the pole shoes. Currently, a high-field magnet is going to be installed at the neutron source BER II of the Helmholtz-Zentrum Berlin at a dedicated instrument for extreme environments ExED [16]. It is expected that continuous fields above 30 T can be realised that allow entering new scientific areas such as the investigation of the fractional quantum Hall effect. In combination with the time-of-flight neutron instrument, both elastic and inelastic experiments under those extreme conditions will be feasible.

It is a lot easier to produce pulsed fields. The difficulty of using these devices for neutrons is related to the total counting statistics. The integrated acquisition time for a given field strength is limited by the pulse duration and the number of repetitions. It has only recently been demonstrated that despite these limitations neutron diffraction patterns can be obtained with pulsed magnetic fields up to 30 T [17].

5.2.4 Electric Fields

High electric fields are comparatively simple to realise for the investigation of structural and dynamical properties of ferroelectrics, piezoelectrics, multiferroics etc. The maximum field strength usually depends on the sample itself since defects or irregular shape may cause electrical breakdown. Since the sample volume must be sufficiently large in order to yield enough counting statistics, the field are usually limited to some 10 kV/cm. In contrast to X-ray scattering experiments, the electrodes itself are almost transparent for neutrons and provide no severe problems.

5.2.5 Chemical Potential: Partial Pressure

Studies of non-stochiometric systems like oxides in solid oxide fuel cells, in mixed valence systems or other oxygen conductors require the control of the oxygen partial pressure in order to avoid changing concentrations [18]. Moreover, chemical loading of zeolites, hydrogen loading of energy materials, gas adsorption, or the investigation of catalytic reaction on surfaces needs well defined in-situ gas handling systems in order to obtain reliable results. At the Helmholtz Centre Berlin a dedicated gas-handling system was installed that can be used for experiments at different neutron instruments [19]. This equipment opens new fields of chemical applications, in particular.

Fig. 5.1 Quasielastic
scattering from hydrogen
during the hydration of
cement [20]

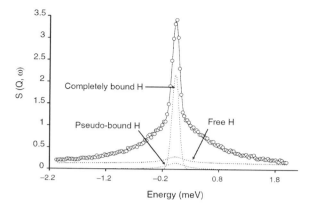

5.3 Single-Shot Kinetic Experiments

We will in the following give a few examples of single-shot experiments. In a
pump-probe configuration the evolution of the sample is monitored after an initial perturbation. Quenching or fast heating is frequently used to initiate transitions
to a not-equilibrium state of matter. Neutron scattering then allows following the
relaxation of this state. In the case of a chemical reaction the initial state is a complex function of reactants and reaction conditions.

5.3.1 Cement Hydration

One important example that has been extensively studied by time-resolved
neutron scattering is the hydration of cement. The extraordinary strong incoherent scattering cross section of hydrogen allows studying the single particle motion of hydrogen in great detail. The form of the quasielastic scattering
from hydrogen provides direct information about its mobility. As shown by
Peterson [20], the scattering profile consists of several distinguishable components reflecting the different mobility of free and bound water. An example is
displayed in Fig. 5.1.

Free water exhibits a large diffusivity and, hence, a broad quasi-elastic spectrum while bound water is much less mobile and leads to a narrow and almost
elastic component. Quantitative interpretation of those data yields the so-called
bound water index BWI that can be used to describe the hydration process. It
could be shown, that the time-evolution of the BWI depends on the polymorph
of tricalcium silicate that is used. Figure 5.2 shows that the diffusion-controlled
hydration is more pronounced in the monoclinic modification resulting in a considerably larger product formation [20, 21].

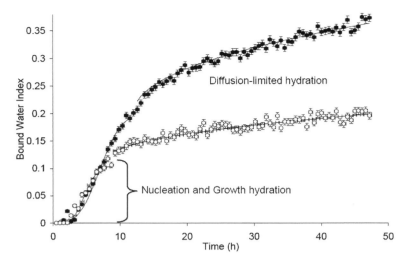

Fig. 5.2 The time-evolution of the BWI for triclinic T_1 (*lower curve*) and monoclinic M_3 (*upper curve*) forms of tricalcium silicate, shown as points with standard deviation. The combined modified Avrami-type nucleation and growth and Fujii and Kondo-derived diffusion-limited models are shown as bold lines [21]

Different kinetic models have been applied to describe the kinetic behaviour quantitatively and, hence the underlying mechanism could be elucidated. Details of the data treatment can be found in [21] or [22]. Additives such as NaOH, $CaCl_2$ or sucrose are found to have a considerable influence on the formation of reaction products [23, 24]. $CaCl_2$, in particular, enhances the hydration of cement as demonstrated by the strong increase of the narrow quasi-elastic component associated with bound water.

The formation of $Ca(OH)_2$ during the hydration reaction of cement can also be monitored by the phonon density of states. Due to the large incoherent scattering cross section of hydrogen, the lattice vibration become more and more visible the larger the $Ca(OH)_2$-content.

5.3.2 Transition Between Ice-Phases

Ice is one of the most fascinating systems offering a wealth of structural features including amorphous and crystalline phases. It is the flexibility of the hydrogen-bond network interconnecting the water molecules that allows for this structural variety. While it is well known that the normal phase of ice exhibits hexagonal symmetry (Ih), up to now fifteen stable and metastable crystalline structures have been observed with interesting kinetic properties. The new high-pressure phase called ice XV has been detected at low temperature only quite recently

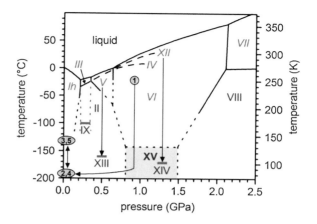

Fig. 5.3 Phase diagram of water [25]

[25]. In Fig. 5.3, the phase diagram of the known crystalline phases is displayed. Neutron diffraction is particularly suited for elucidating the structural details as well as the transitions and, hence, a good number of neutron-related publications are dealing with the study of the water phase diagram as a function of temperature and pressure.

In a recent investigation, Hansen et al. could demonstrate that cubic ice Ic is formed if recovered high-pressure phases of ice IX or ice V are annealed at temperatures near 145 K [26]. Careful Rietveld refinement of the diffraction data allowed performing a quantitative phase analysis. It is shown that the formation of cubic ice takes some hours to proceed. Being metastable, the cubic phase is subsequently transformed into hexagonal ice on heating. The corresponding time-evolution of the stacking sequence could be determined in detail. Figure 5.4 shows the variation of stacking probabilities during the transition from cubic to hexagonal ice.

In addition to the crystalline ice forms, several amorphous phases with different densities are observed. Due to the rather small difference in the chemical potential of these phases, the kinetics of the transition between them is often rather sluggish. Koza et al. give a review about the transformations between different amorphous structures of deuterated ice [27]. As an example, Fig. 5.5 shows the time-evolution of the neutron diffraction pattern during the transition from high- to low-density amorphous ice.

The broad patterns are characteristic for non-crystalline structures and the shift of the intensity maxima to smaller scattering vectors Q directly reflects the decrease of the packing density of the water molecules. It is also seen from Fig. 5.5 that there is a transient increase in intensity at small Q. Small angle scattering experiments helped to study this feature in more detail (see Fig. 5.6).

Obviously, the scattered intensity near 0.1 $Å^{-1}$ initially starts to increase (upper part of the figure) reaching a maximum and finally decreases towards the final state (lower part). In contrast to this finding, the intensity of the very low-Q regime decreases monotonically in time. This so-called Porod-regime is reflecting the grain

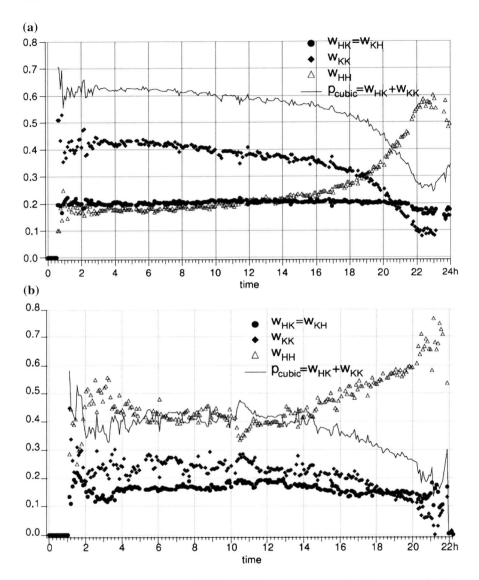

Fig. 5.4 Evolution of the proportion of cubic stacking sequences over time ($w_{HK} + w_{KK}$), of the probability of finding pairs of similar stacking (w_{KK} and w_{HH}) and of interfaces between different stacking types (w_{HH}) for ice Ic as produced from ice V (**a**) and ice IX (**b**) [26]

structure of the sample, while the intermediate regime provides information about the heterogeneity of the sample. On the basis of these observations it is claimed that the transition between the amorphous ice phases occurs on two different time-scales: The formation of an intermediate structure characterized by a strong heterogeneity is rather sluggish and depends on the history and the preparation of the

Fig. 5.5 Time evolution of the neutron diffraction pattern during the transition from high to low density amorphous ice [27]

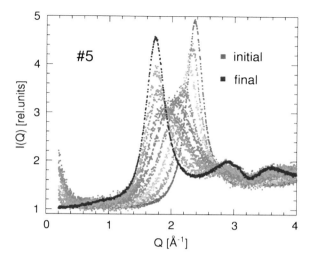

Fig. 5.6 Evolution of the small angle intensity during the transformation from high- to low-density amorphous ice [27]

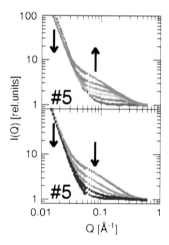

sample under consideration. Typical time scales are of the order of 100 min. The subsequent step of creating a homogenous low-density state is much faster and completed after several minutes for each of the different samples investigated.

5.3.3 Crystallization of Glass-Phases

Another actual topic of kinetic studies using neutrons is the crystallization of glass phases upon annealing. Recently, Ma et al. could show that time-resolved diffraction on a time scale of minutes allows extracting the evolution of the atomic

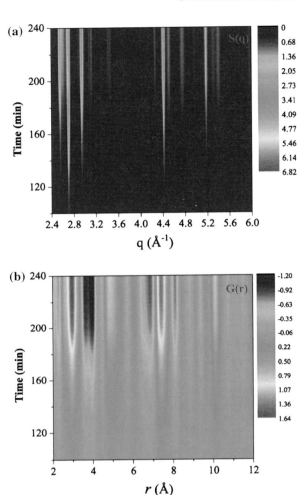

Fig. 5.7 Time evolution of the total structure factor S(q) (*top*) and its Fourier transform, the pair distribution function G(r) (*bottom*) during the crystallization reaction of $Zr_{50.4}Cu_{40.6}Al_9$ metallic glass at 420 °C [28]

pair distribution function for a Zr-based metallic glass [28]. The upper part of Fig. 5.7 shows the time evolution of the diffraction pattern. While at the beginning the characteristic diffuse scattering of the glass phase is hardly seen in this contour plot, the appearance of well-defined Bragg reflections of the crystalline phase dominates the intensity map after about 2 h. From the Fourier-analysis of these data, the time-dependence of the pair distribution function is obtained as shown in the lower part of Fig. 5.7, which illustrates impressively the evolution of the ordered structure.

The quantitative interpretation of the experimental results allows extracting the time dependence of the crystalline volume fraction. It could be shown that this quantity is well described by a stretched exponential function according to the Johnson–Mehl–Avrami model. It is inferred that the underlying process is interface-controlled.

5.3.4 Crystallization Reactions and Extreme Conditions

Reactions under hydrothermal conditions require a sophisticated sample environment. Recently, first results with a newly designed hydrothermal cell developed at the Australian Nuclear Science and Technology Organisation ANSTO for use with the powder diffractometer WOMBAT have been published by Xia et al. [29]. Due to the fact that a Ti-Zr alloys with zero coherent scattering length has been used as cell-material excellent background conditions could be achieved. At the same time, a stainless-steel mesh supporting the sample serves as an internal standard. One of the first applications of this cell addressed the kinetics of the transformation from leucite ($KAlSi_2O_6$) to analcime ($NaAlSi_2O_6 \cdot H_2O$) as an example of pseudomorphic transformations, Fig. 5.8.

It was known that under hydrothermal conditions leucite is dissolved in an aqueous NaCl-solution followed by the precipitation of analcime and the overall reaction can be summarized as

$$KAlSi_2O_6 + Na^+ + H_2O \rightarrow NaAlSi_2O_6 \cdot H_2O + K^+.$$

The time scale of this process was, however, unknown. The time-evolution of the corresponding diffraction pattern at 210 °C is shown in Fig. 5.8. The individual peaks can be well assigned to the different phases and a Rietveld-least squares fitting procedure yields the time-dependence of the respective mass fractions. These data are displayed in Fig. 5.9 and can be well described by an Avrami-equation with an exponent of 2.

5.3.5 Self-Assembly in Soft Matter

Processes of self-assembly in soft matter systems can be well monitored by real-time small angle neutron scattering since the scattering contrast can be adjusted by selective deuteration of one component. An illustrative example is the pressure-induced reorganisation of micellar solutions. In aqueous solution, surfactant molecules of tetradecyl-trimethylammonium-bromide (TTAB) are forming almost spherical micelles as long as the critical micelle concentration (cmc) of 3.8 mM is exceeded. The self-organisation of these particles is reflected by a correlation peak at wave vector transfers close to 0.05 Å^{-1} as shown in Fig. 5.10a [30] for a 25 mM solution of TTAB. Using specific models one is able to extract not only the geometrical parameters of the micelles but also their charges.

If pressure is applied to this system, only minor structural changes are observed up to 1000 bar and can be attributed to the compressibility of the solvent (D_2O). After increasing the pressure to 1100 bar, however, the scattered intensity drops dramatically on a time scale of hours due to the beginning crystallization of the system and the formation of so-called hydrated crystals as illustrated by the sequence of scattering profiles in Fig. 5.10b.

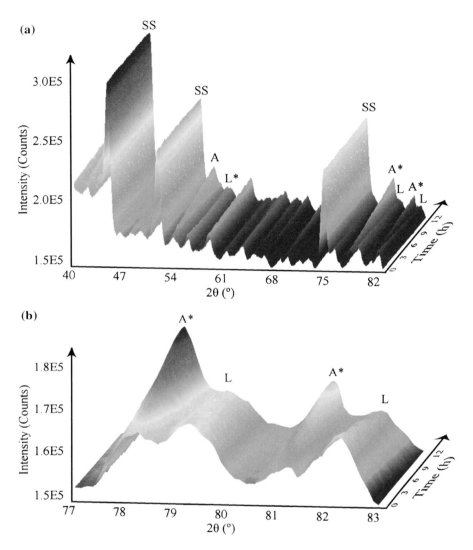

Fig. 5.8 In-situ neutron diffraction patterns of the transformation from leucite ($KAlSi_2O_6$) to analcime ($NaAlSi_2O_6 \cdot H_2O$). **a** Two-theta range from $40°$ to $83°$, showing the progressive phase transformation. **b** Two-theta range from $77°$ to $83°$, highlighting the progressive increase of analcime peaks and decrease of leucite peaks. *L* leucite peaks, *A* analcime peaks, *L*(A*)* overlapped peaks where leucite (analcime) is the dominant phase, *SS* stainless steel peaks [29]

This simple behaviour changes drastically if the concentration is increased to, say, 360 mM. Here, the transition starts at smaller pressures but it is not completed any more. Rather, the time-resolved experiments show that even after long periods of time a considerable part of the correlation peak, which is a signature of the liquid micellar phase, remains detectable and a new equilibrium

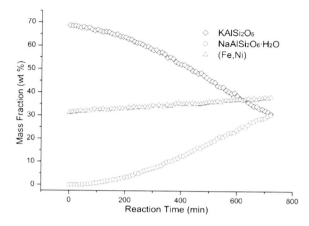

Fig. 5.9 Mass fractions of the three involved phases as a function of reaction time during the transformation from leucite ($KAlSi_2O_6$) to analcime ($NaAlSi_2O_6 \cdot H_2O$) [29]. Note that the total mass of the sample changed from 1.25 g at the beginning of the reaction to 0.92 g at the end because of the loss of small particles from the mesh tube; however, the mass of the stainless steel mesh tube remained constant at 0.57 g, so the mass fraction of (Fe,Ni) slightly increased during the process, serving as an excellent internal standard for quantification

state is formed. For a 360 mM solution of TTAB in D_2O the time evolution of the small angle scattering intensity after applying a hydrostatic pressure of 700 bar is displayed in Fig. 5.11.

The correlation peak shifts towards smaller scattering angles and simultaneously the intensity decreases. The time-dependence of the integrated intensity (Porod-invariant) is shown in the inset along with the behaviour at 900 bar. Due to the larger thermodynamic driving force at higher pressures the kinetics become faster and the residual intensity corresponding to the remaining liquid micellar phase becomes smaller. In a pressure-concentration phase diagram, a two-phase regime develops towards higher concentrations where the solid and the liquid micellar phase coexist.

Due to the sluggish kinetics of the underlying process a careful determination of the phase diagram needs the time-resolved data in order to extrapolate the saturation value of the intensity.

While in this system, the correlation peak in the small angle regime is the characteristic feature for the (homogeneous) micellar solution, which is reduced by the fractional crystallization, the reverse feature, the growth of the intensity maximum is frequently observed, if inhomogeneous structures are formed by demixing reactions.

Exchange reactions of single surfactant molecules between micelles and the solution can be studied using the unique possibility of contrast variation by deuteration. Lund has shown that block-co-polymer micelles built from poly(ethylene-propylene) PEP and poly(ethylene oxide) PEO in water or water dimethylformamide (DMF) mixtures exhibit a complex kinetic behaviour [31]. The exchange of individual molecules between different micelles was monitored

Fig. 5.10 **a** Small angle scattering from 25 mM TTAB-solution at 25 °C at different pressures. **b** Time evolution of scattering profiles at 1100 bar (in the inset the time dependence of the Porod invariant is shown). Note, that the individual curves are shifted vertically for clarity [30]

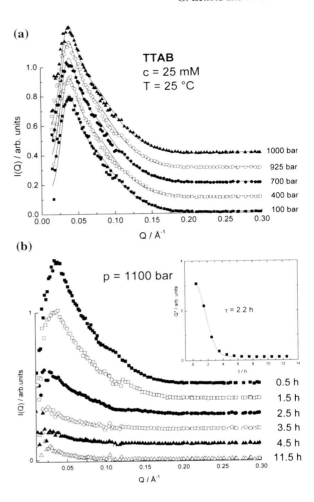

by mixing deuterated and hydrogenated polymers. The solvent was chosen to yield just the same scattering length density as a micelle that consists of a random mixture of both types of polymers. Hence, those micelles are invisible in small angle neutron scattering experiments. If, however, the micelles consist exclusively of deuterated or hydrogenated polymers the scattering contrast with respect to the solvent is large and the size and the shape of micelles, e.g., can be determined with accuracy from the scattering profiles. In a kinetic experiment, Lund et al. have prepared both types of micelles separately. After mixing both systems individual polymer molecules may leave one micelle and migrate to another, thus leading to a more random isotope distribution. Hence, the scattered intensity will decrease in time and the rate of the exchange reaction can be determined [32].

Using pure water as solvent, no exchange reaction was observed over hours. Adding dimethyl-formamide, however, leads to a significant reduction of

Fig. 5.11 Time evolution of SANS intensity of 360 mM TTAB solution at 25 °C and 700 bar. In the inset, the time dependence of the Porod invariant is shown— for comparison the data for 900 bar are also included [30]

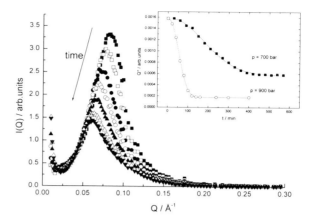

the interfacial tension and, consequently, also to an increased expulsion and exchange rate of polymers from and between the micelles. The corresponding relaxation function, i.e. the square root of the normalized intensity, is shown in Fig. 5.12 for a system of PEP1-PEO20 micelles with a volume fraction of 1 % in a 25 % DMF-water mixture at different temperatures. It can be seen that the kinetics is much more complicated than expected from the simple theory of Halperin and Alexander [33] that predicts a simple exponential decay. It could be excluded that the deviations are due to polydispersity. Rather, the data suggest that only part of the polymer molecules are able to escape rapidly from the micelle—these are molecules that are located close to the interface and have a compact conformation. Other molecules in the interior of the micelle suffer from topological interactions with the surrounding polymers and dynamic constraints that lead to a much slower exchange kinetics. Therefore a broad range of relaxation times may be expected that leads to the kinetic behaviour as reflected by Fig. 5.12.

Small angle neutron scattering is also suitable for the study of the formation and growth of larger aggregates like vesicles. Grillo et al. [34] have investigated the transformation from micelles to vesicles of sodium bis(2-ethyl hexyl)sulfosuccinate (AOT) induced by the addition of salts. Using a stopped flow apparatus, the time-evolution of the small angle intensity profile could be observed from 0.5 s up to several hours after mixing. The results are shown in Fig. 5.13. Quantitative interpretation of the data in terms of the structure factors of vesicles allows extracting the time-dependence of the vesicle radius.

As demonstrated by the log–log plot in Fig. 5.14 the radius grows as a power law with an exponent of 1/6, and this over more than four decades in time. This finding indicates that the aggregation rate is essentially determined by the diffusion of micelles that are incorporated within already existing vesicles.

On an even larger length scale the gelation of milk has recently been studied by spin-echo and ultra small angle scattering (SESANS and USANS) [35]. The formation of yogurt from milk can be described as an aggregation process

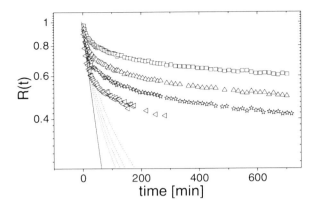

Fig. 5.12 Relaxation kinetics of the PEP1-PEO20 micelles in a 25 mol % DMF/water mixture system at a total volume fraction of 1 % and different temperatures in a semi-logarithmical presentation. From *top to bottom*: 47, 55, 60, and 65 °C. The *solid line* displays the theoretically expected single exponential decay. The *dotted lines* indicate the expected theoretical decay taking into account the polydispersity of the core blocks [32]

Fig. 5.13 Time evolution of small angle scattering intensity after mixing a 0.75 wt. % AOT-solution with 0.042 M NaCl [34]

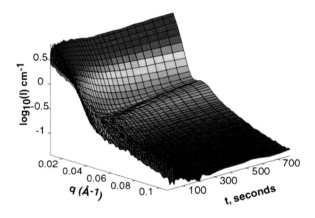

of casein-micelles under the influence of bacteria, increased acidity etc., thus creating a 3D network with increasing length scales. In concentrated real systems, the evolution of these structures cannot be observed by light-scattering or microscopy, since they are in general opaque. Hence, small angle neutron scattering is the method of choice for a detailed study of the structural properties. Advanced techniques like spin-scho small angle neutron scattering SESANS or ultra small angle neutron scattering USANS are, however, needed in order to access characteristic length scales of the order of several 100 nm. Figure 5.15 shows that on a time scale of hours the gelation takes place leading to a network of μm-sized structures.

Fig. 5.14 Time-dependence of the vesicle radius in a log–log representation for 0.75 wt. % AOT in 0.042 M NaCl [34]

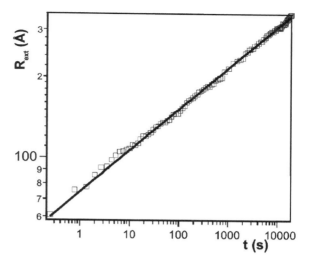

Fig. 5.15 Time dependence of the characteristic length scale for the formation of yogurt as obtained from USANS-experiments [35]

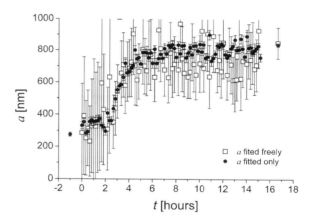

5.4 Stroboscopic Kinetic Experiments

If the time scale of kinetic processes is comparable to or even less than character-istic counting times needed to obtain sufficient statistics, single shot experiments are no longer feasible. In this case, stroboscopic or pump-and-probe techniques can be used where the sample is cycled between to different thermodynamic states. This needs the periodic perturbation of the sample and an appropriate data acquisition system that counts neutron not only as a function of instrument param-eters like Q or E but also as a function of time with respect to the perturbation. Time scales down to the microsecond regime are accessible in this way even for real-time inelastic scattering experiments. If the process under consideration is repeatable and the scattered intensity is accumulated over a large number of cycles detailed information about the changes of dynamics and chemical bonding in condensed matter can thus be obtained. Some selected examples of those studies will be presented in the following.

5.4.1 Decomposition Kinetics

It is well known for metallic alloy systems that demixing after thermal heat treatment and subsequent quenching leads to the formation of precipitates that improve the mechanical properties of the material (precipitation hardening). The structure and morphology of precipitates depends on the type of heat treatment and the kinetics of the decomposition process. Two different mechanisms can be distinguished in order to describe the phase separation process: On one hand nucleation and growth is based on the statistical formation of small nuclei that happen to possess the concentration of the product phases. These nuclei then grow in time. Spinodal decomposition, on the other hand, starts from tiny concentration fluctuations that are spatially correlated within the homogeneous phase and grow in amplitude. This mechanism is dominating if the phase separation starts from a thermodynamically unstable state where even small concentration fluctuations are favoured over the homogeneous state. Being a diffusion-limited process, there is a preferred wavelength of growing fluctuations that leads to a correlation peak in the small angle scattering regime. In a second stage, coarsening or Ostwald ripening takes place, associated with an increase of the wavelength and the shift of the correlation peak to smaller wave numbers.

While spinodal decomposition has been studied in some detail in alloys and polymer blends in the past, ionic systems are hardly investigated. Recently, it could be shown, however, that quasi-binary systems of the silver-alkali halide type are almost ideal model systems for detailed investigations of demixing mechanisms and, in particular the competition between nucleation and fluctuations [36–40]. It could be shown, that time resolved diffraction, small angle scattering as well as inelastic scattering from phonons provide complementary information about these systems and allow distinguishing between different aspects on different time-scales. It is observed that the splitting of Bragg reflections due to the difference of the lattice parameters of the product phases is almost independent of the intrinsic chemical demixing process and occurs on a rather long time-scale of hours to months. In contrast, small angle scattering clearly shows the growth of the correlation peaks within minutes. As an example, the evolution of a Bragg reflexion in $Ag_{1-x}Na_xCl$ is shown in Fig. 5.16a after quenching a mixed crystal from the homogeneous phase at 623 K deep into the demixing region below the coherent spinodal line. There is almost no change of the profile even after a few hours apart from wings that are attributed to the concentration modulation [38]. The equilibrium structure with two well

Fig. 5.16 **a** Time evolution of the (200)-Bragg reflexion of $Ag_{0.5}Na_{0.5}Cl$ at 373 K. **b** Bragg profile of a $Ag_{0.4}Na_{0.6}Cl$ single crystal after ageing for more than one year at ambient temperature [39]

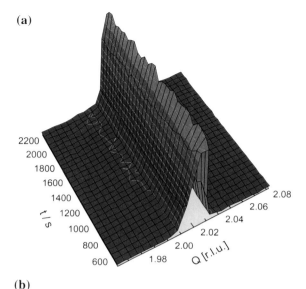

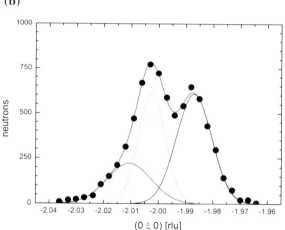

separated Bragg peaks for the two product phases is, however, not attained even after months as illustrated in Fig. 5.16b.

The corresponding small angle scattering results are shown in Fig. 5.17 and reveal that the characteristic correlation peak starts to grow within the first few minutes after the quench. Unlike the behaviour of linear systems where the peak position remains constant, in $Ag_{0.5}Na_{0.5}Cl$ the peak is shifted towards smaller Q already at small ageing times.

The inset shows the time-dependence of the peak position: For longer times, the coarsening behaviour Q_{max} proportional to $t^{-1/3}$ according the law of Lifshitz-Slyozov is obtained. Note, that the integrated intensity (in three-dimensional

Fig. 5.17 Time-evolution of the small-angle intensity of Ag$_{0.5}$Na$_{0.5}$Cl at 373 K (inset shows the time-dependence of the peak position) [37]

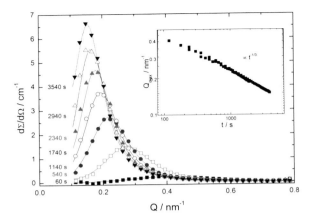

Fig. 5.18 Time-evolution of the TA$_1$[0.2 0.2 0] phonon in Ag$_{0.45}$Na$_{0.55}$Br during demixing at 373 K. The lines are the results of a fit with two Lorentzians, except for the spectrum at t = 45 s which corresponds to the homogeneous state [39]

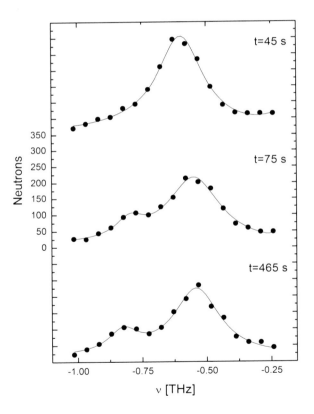

Q-space) is almost constant over the whole time interval proving evidence that the final concentrations are reached already after some minutes.

Inelastic scattering from phonons in single crystals has been used to monitor the variation of the local interionic interactions directly. Due to the

Fig. 5.19 Time-dependence
of the autocorrelation
function for phonons and
Bragg reflections (*top*),
integrated small angle
intensity (*middle*) and
position of correlation peak
during demixing at 373 K
[39]

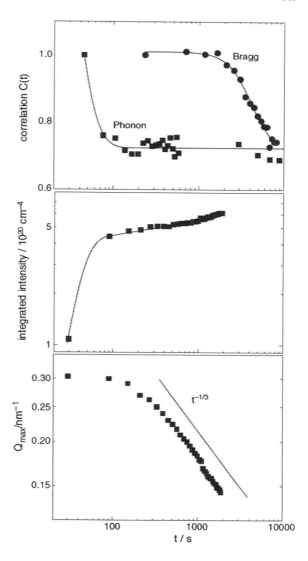

different elastic constants of the constituents, acoustic phonons separate into
two distinct components during the demixing reaction. For a single crystal with the overall composition $Ag_{0.45}Na_{0.55}Br$, the time-evolution of phonon spectra is shown in Fig. 5.18. It is clearly seen, that the phonons separate
within some minutes. This is the most direct proof that the demixing reaction
is, in fact, a rather fast process.

The different time-dependent behaviours of Bragg-reflections, phonons,
small-angle intensity and correlation peak position are compared in Fig. 5.19
for $Ag_{0.45}Na_{0.55}Br$. It is readily seen that phonons and small angle scattering go

hand in hand while the Bragg peaks are delayed. Reason for this behaviour is the existence of coherence strains that keep the lattice almost invariant during the chemical demixing. Only if the precipitates are grown beyond a characteristic size of about 50 nm (as estimated form the position of the correlation peak), the lattice starts to relax giving rise to the splitting of Bragg peaks.

In the corresponding chlorine system, the coherence strains are considerably smaller and so is the driving force for lattice relaxation, which is still not completed after months as shown in Fig. 5.16.

Spinodal phenomena are also observed in crystallization studies of fluoroapatite glass–ceramics. These systems are particularly interesting due to their medical applications and also for opto-electronical applications. It is believed that on cooling cast calcium-fluorapatite glasses are entering a two phase region and an amorphous phase separation takes place leading to a coexistence of a Ca- and F/P-rich phase, on the one hand, and an Al/Si-rich phase, on the other hand, with different glass temperatures. Depending on the initial composition, the phase separation is either dominated by nucleation or spinodal decomposition. In the latter case, the characteristic features of coherent concentration fluctuations can be observed in small angle scattering experiments [41]. On annealing at temperatures close to the respective glass transition temperature, crystallization of the respective phases takes place. This can be monitored by the variation of the small angle intensity profiles as shown in Fig. 5.20.

It can be seen that the correlation peak starts to grow and shifts slightly to smaller momentum transfer. At the same time, the intensity at very small q increases drastically due to the incipient formation of the crystalline phase. At longer ageing times, the correlation peak is completely drowned by the low-q contribution, which seems to saturate after about 800 s. This is an indication that the size of the crystals that can be estimated from the width of the intensity distribution remains in the nanometer regime due to the presence of the second phase that was built during the amorphous phase separation.

It has been pointed out by Hill et al. [41] that the use of neutron techniques is essential for these experiments since they provide the intrinsic bulk behaviour while earlier X-ray data were strongly affected by surface effects. Here, nucleation or phase separation processes may be different to the bulk and, moreover, the surface composition might vary due to the loss of volatile species like silicon tetrafluoride.

5.4.2 Switching Processes In Ferroelectrics and Piezoelectrics

Using stroboscopic thermal neutron diffraction or inelastic scattering, the switching processes or domain redistribution in ferroics under the influence of e.g. pulsed electric fields can be characterized on a microscopic scale. In most cases, the rise time of the electric fields is limiting the time-resolution, due to the inherent capacities present in the sample, leads and plugs. Nevertheless, real-time data in the microsecond-regime could be achieved for several systems.

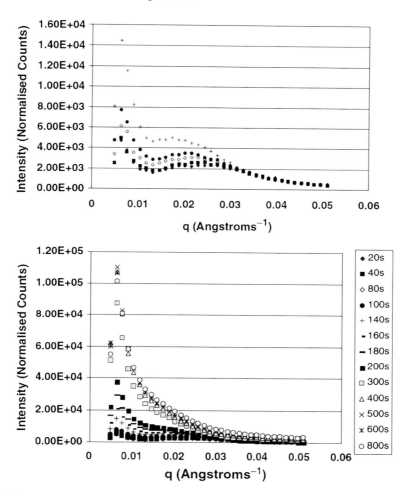

Fig. 5.20 Time-evolution of small angle neutron scattering from a calcium-fluoroapatite glass during annealing at 750 °C [41]

One example is the field-induced polarization inversion in ferroelectric PZN ($PbZr_{1/3}Nb_{2/3}O_3$)–PT ($PbTiO_3$) single crystals. Under zero field conditions, the spontaneous polarization can be oriented along one of the eight (111)—directions of the pseudo-cubic elementary cell. Hence, eight different domains can be distinguished that simultaneously exhibit different piezoelectric strains and slightly different positions of the corresponding Bragg reflections in reciprocal space. Application of an external field will favour one or more of these domains over the others. The redistribution can be monitored by observing the changes of the Bragg-profiles. In Fig. 5.21, selected data are shown for a PZT-PT crystal at different static electric fields as obtained by Daniels et al. [42]. The expected Bragg-positions for the individual domains are marked by black squares.

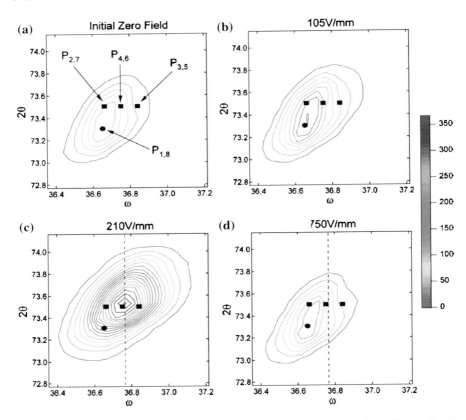

Fig. 5.21 Profiles of the $(222)_c$-Bragg reflection of a PZN-PT single crystal at different electric fields [42]

During polarization reversal under the influence of a bipolar 750 V/mm square wave, the intensity variation was measured at constant crystal orientation along the dashed line. The result is shown in Fig. 5.22. It is clearly seen, that the polarization reversal is associated with an increase of the peak-intensity and occurs almost instantaneously on the time-scale of seconds. There is, however a long time-relaxation with a characteristic decay time of about 5 s leading to a significantly reduced intensity. It is argued that this effect is due to the creep nature of the underlying domain structure [42].

Not only can the domain distribution in ferroelectrics be influenced by an external electric field. Moreover, the ferroelectric transition itself can be induced and its kinetics can be investigated by time-resolved neutron scattering. Modulated ferroelectrics of the A_2BX_4-type (A = K, Rb, NH$_4$,..., B = Zn, Se, ..., X = Cl, Br, O,...) e.g., have been studied in some detail in the past [43–45]. This family of compounds exhibits a variety of modulated phases and the prototypic sequence of phase transitions consists of the high-temperature para-phase, an incommensurate (INC) phase and a low-temperature ordered and sometimes ferroelectric commensurate (C) phase. In this case, the C-phase is stabilised by an electric field and, hence, the lock-in transition

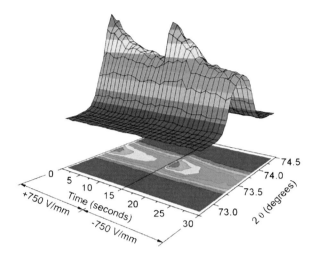

Fig. 5.22 Time evolution of the Bragg-profile during a bipolar square wave of amplitude 750 V/mm and period 30 s [42]

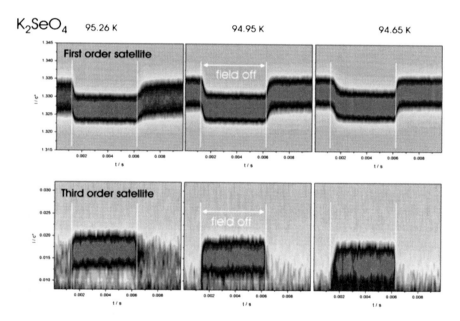

Fig. 5.23 Time-evolution of the first order (*top*) and the third order (*bottom*) satellite patterns in K_2SeO_4 at three temperatures close to the lock-in transition [40]

between the INC and the C-phase may be driven by the external field. The underlying mechanism is governed by the nucleation and growth of topological defects called stripples and antistripples [46, 47]. Details can be studied, if the characteristic satellite reflections are observed experimentally during the transition. For the case of K_2SeO_4, the time-evolution of the intensity maps is shown in Fig. 5.23 not only of the first order satellite (top) but also of the third order one (bottom) for three different temperatures [48].

Since in the C-phase the third order satellite coincides with the fundamental Bragg peak, the switching between the phases is most clearly reflected by the appearance of this new peak when entering the INC-phase. The sequence of temperatures demonstrates that the INC wave vector approaches gradually the commensurate value of 2/3 c^* on cooling. Careful inspection of the time-evolution of both types of satellites reveals that the relaxation time is considerably temperature dependent. The formation of the ferroelectric phase is delayed by about 200 μs at higher temperatures, while the back-transformation into the INC-phase becomes more sluggish at lower temperatures. The varying thermodynamic driving forces can explain this behaviour.

Even if the phase transitions in K_2SeO_4 are associated with a soft phonon and phason mode, their frequencies close to the lock-in transition are not at all affected by an applied electric field. It is believed that the INC-phase consists of an ordered arrangement of nanometer sized commensurate domains that grow during the (field-induced) lock-in transition. Hence, the local dynamics is already determined by the commensurate structure and independent of the size of domains.

Of particular interest are systems in which the external field is able to affect the balance of interatomic forces, thus inducing softening and a phonon-mediated phase transition. $SrTiO_3$ as a member of the perovskite family is a famous example. Unlike its isotypical counterpart $BaTiO_3$, it does not become ferroelectric even at the lowest temperatures. The transition behaviour is governed by the competition of two soft phonon modes: A zone-boundary phonon becomes soft and on condensation at 105 K, it leads to an antiferrodistortive orthorhombic phase [49]. The zone-centre polar softmode, on the other hand, exhibits an incomplete softening and remains stable down to the lowest temperatures. According to Müller et al. [50], quantum fluctuations prevent its condensation and, hence, the transition to a ferroelectric phase. The application of a sufficiently strong external electric field, however, can induce the ferroelectric order. It has been shown by Eckold [40] that, the softmode itself is altered by the electric field thus leading to a complete softening at the transition and subsequently to a stabilisation within the polar phase. The kinetics of this transition can thus be characterised on a microscopic scale by observing the time-evolution of phonon spectra during the application of pulsed electric fields. Figure 5.24 shows first results obtained with the three-axes spectrometer PUMA at the FRM-II neutron source.

The variation of the inelastically scattered intensity at the Gamma-point (002) is presented (Fig. 5.24a) in the form of a contour plot as a function of frequency and real-time during a 200 Hz pulsed electric field of 8 kV cm^{-1}. Note, that the strong elastic Bragg peak at zero frequency is omitted for clarity. If the field is off, a maximum of the intensity is observed at about 0.75 THz. After switching the field on the maximum is shifted to higher frequencies of about 1 THz as soon as the transition has taken place. As can be seen from the contour plots the phonons are heavily damped. Hence, a careful analysis is needed in order to obtain reliable phonon frequencies. Taking into account the non-linear dispersion of the softmode, contamination from acoustic modes as well as the 4D-resolution function of the three-axes

Fig. 5.24 Time evolution
of the Γ-point soft-mode in
SrTiO₃ at 30 K under the
influence of a pulsed electric
field of 8 kV/cm. **a** Intensity
distribution at Q = (2 0 0).
Note, that the Bragg-intensity
around ν = 0 is omitted.
The field is on within the
time-interval marked by the
vertical lines. **b** Variation
of the phonon frequency
as obtained by fitting the
intensity distribution with a
damped harmonic oscillator
[40]

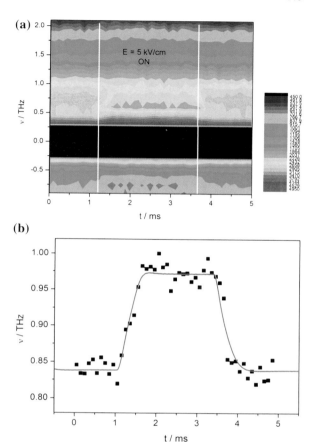

spectrometer, we were able to determine the time-dependence of the softmode frequency. The results are shown in Fig. 5.24b. Obviously, the phonon reacts on a time scale of microseconds and seems to follow the applied field instantaneously. Domain walls or other defects do not significantly disturb the dynamical behaviour.

These data show for the first time that it is possible to characterise transformation processes by real-time phonon spectroscopy even on time scales down to microseconds. This technique provides the most direct information about the varying interatomic interactions and chemical bonding.

5.4.3 Limitations of Stroboscopic Techniques

There is a principle limitation of the time scales accessible with stroboscopic neutron scattering due to the energy or wavelength distribution of the neutrons. If neutrons with different energies are scattered at a given moment by the sample they will not arrive simultaneously at the detector. Thermal neutrons with typical velocities of 4000 ms^{-1} (wavelength 1 Å) need e.g. 0.5 ms to travel a distance of 2 m between sample and detector. If there is a velocity or wavelength spread of 3 % neutrons that leave the sample simultaneously will be registered in the detector over a time-interval of 15 μs. This gives a lower limit for the attainable time-resolution in kinetic experiments. For cold neutrons this limitation becomes more serious. Typical small angle experiments use neutrons with wavelengths around, say, 5 Å with 10 % bandwidth. If the distance between sample and detector is chosen to be 10 m, the time-resolution becomes worse than 1 ms. In order to overcome this limitation, Wiedenmann et al. have proposed a pulsed beam technique called TISANE [51]. In this approach a chopper placed at a certain distance before the sample generates a bunch of neutrons with a rather large wavelength spread. Neutrons with different wavelengths hit the sample at different times and probe a different state if the sample is perturbed periodically. If scattered elastically, these neutrons can be distinguished by their respective arrival time at the detector. Moreover, the careful synchronization of chopper period, sample perturbation and detector acquisition guarantees that corresponding neutrons of different pulses are always counted in the same time frames of the detector. Hence, neutron counts can be accumulated over a large number of cycles until sufficient statistics is obtained. This method allows accessing time-scales well below 1 ms in small angle experiments even if slow neutrons are employed.

5.4.4 Magnetic Relaxation in Ferrofluids

One of the first examples for the application of this method was the investigation of the structural relaxation of ferrofluids under the influence of AC magnetic fields. Polarized neutrons were used to monitor the spatial orientation distribution of magnetic moments and their variation with the applied magnetic field.

While without any field, the scattered intensity is isotropic and it exhibits a significant angular dependence if a magnetic field is switched on. In Fig. 5.25, the time-dependent intensities scattered from a cobalt-ferrofluid are shown for wave vectors parallel and perpendicular to the applied field. In the upper part of this figure, the results obtained with conventional stroboscopic technique are presented for different frequencies of the applied field. As a reference, the isotropic intensity level for zero field is shown by the dashed line.

Clearly, the time-dependent contrast is almost vanishing at a frequency of 400 Hz. Due to the non-zero wavelength band, the time resolution of this experiment is limited to 1 ms and, hence the signal is averaged over almost half of the period at this frequency. Quantitative analysis has shown that the decay of the signal is not due to the finite mobility of the magnetic moments but simply to the resolution effect.

Application of the TISANE method allows increasing the frequency beyond 1 kHz without restrictions from resolution as shown in the lower part of Fig. 5.25. Here, the intensity modulation is still visible at 1099 Hz and the number of freely oscillating magnetic moments gives its amplitude. At 1424 Hz the scattering contrast

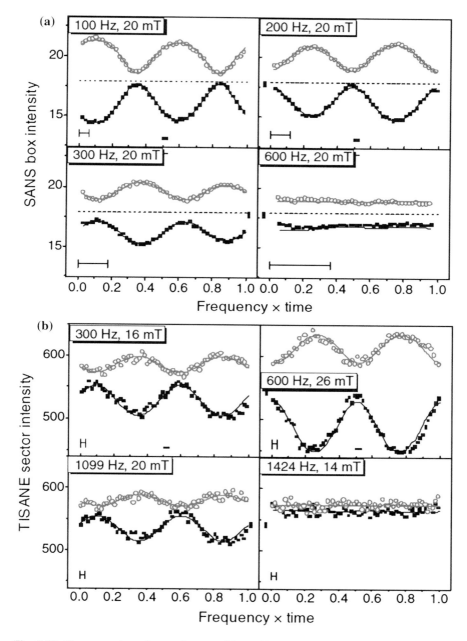

Fig. 5.25 Frequency dependence of scattered intensities from Co-ferrofluid. Bars indicate the experimental resolution. Open (closed) symbols represent the intensity of neutrons scattered into a sector parallel (perpendicular) to the applied field. The *dashed lines* are the results from static SANS experiments without applied field. **a** Stroboscopic SANS mode. **b** TISANE mode. *Solid lines* are the results of fits to a theoretical model [51]

is fading away since the moments are no longer able to follow the external field. Hence, these real-time experiments provide direct information about the relaxation behaviour of the magnetic moments even on time scales well below the millisecond regime.

5.5 Further Prospects of In-situ Investigations

Great progress has been achieved in the field of neutron scattering from materials under extreme or non-equilibrium conditions. Developments for sophisticated sample environment, multiplexing and real-time techniques lead to new fields of application for elastic and inelastic neutron scattering. At different places, instruments are being equipped with user-friendly sample environment. To mention only one specific example: the VULCAN diffractometer at the Spallation Neutron Source SNS in Oak-Ridge focusses on in-situ and real-time experiments for the investigation of the mechanical behaviour of materials [52]. Going one step further many facilities are in the process of setting-up dedicated interface laboratories to allow for more sophisticated experiments by providing extensive off-line preparation and in-situ equipment. Examples are the partnerships for soft condensed matter as well as the extreme conditions initiative at the Institut Laue Langevin in Grenoble, France.

References

1. M. Steiner, J. Low Temp. Phys. **135**, 545 (2004)
2. L. Hennet, S. Krishnan, I. Pozdnyakova, V. Cristiglio, G.J. Cuello, H.E. Fischer, A. Bytchkov, F. Albergamo, D. Zanghi, J.-F. Brun, S. Brassamin, M.-L. Saboungi, Pure Appl. Chem. **79**, 1643 (2007)
3. A. Meyer, S. Stüber, D. Holland-Moritz, O. Heinen, T. Unruh, Phys. Rev. B **77**, 092201 (2008)
4. J. Kozaily, L. Hennet, H.E. Fischer, M. Koza, S. Brassamin, S. Magazù, F. Kargl, Phys. Status Solidi C, 1–4 (2011). doi:10.1002/pssc.201000753
5. S. Klotz, J.M. Besson, G. Hamel, R.J. Nelmes, J.S. Loveday, W.G. Marshall, High Pressure Res. **14**, 249 (1996)
6. Y. Le Godec, M.T. Dove, S.A.T. Redfern, M.G. Tucker, W.G. Marshall, G. Syfosse, S. Klotz, High Pressure Res. **23**, 281 (2003)
7. T. Strässle, S. Klotz, G. Hamel, M.M. Koza, H. Schober, Phys. Rev. Lett. **99**, 175501 (2007)
8. S. Klotz, Th Strässle, A.L. Cornelius, J. Philippe, Th Hansen, Phys. Rev. Lett. **104**, 115501 (2010)
9. Y. Zhao, J. Zhang, H. Xu, K.A. Lokshin, D. He, J. Qian, C. Pantea, L.L. Daemen, S.C. Vogel, Y. Ding, J. Xu, Appl. Phys. A **99**, 585 (2010)
10. E.J. Yearley, L.A. Sasa, C.F. Welch, M.A. Taylor, K.M. Kupcho, R.D. Gilbertdon, R.P. Hjelm, Rev. Sci. Instr. **81**, 045109 (2010)
11. J. Sharma, S. King, L. Bohlin, N. Clarke, Nucl. Instr. Meth. A **620**, 437 (2010)
12. G. Garcés, G. Bruno, A. Wanner, Acta Mater. **55**, 5389 (2007)
13. B. Malard, T. Pirling, K. Inal, E. Patoor, S. Berveiller, Mat. Sci. Forum, Residual Stresses VII, **524–525**, 905 (2006)
14. C.D.M. Liljedahl, O. Zanellato, M.E. Fitzpatrick, J. Lin, L. Edwards, J. Fatigue **32**, 735 (2010)

15. W. Woo, Z. Feng, X.L. Wang, D.W. Brown, B. Clausen, K. An, H. Choo, C.R. Hubbard, S.A. David, Sci. Technol. weld. **12**, 298 (2007)
16. P. Smeibidl, A. Tennant, H. Ehmler, M. Bird, J. Low Temp. Phys. **159**, 402 (2010)
17. S. Yoshii, K. Ohoyama, K. Kurosawa, H. Nojiri, M. Matsuda, P. Frings, F. Duc, B. Vignolle, G.L.J.A. Rikken, L.-P. Regnault, S. Michimura, F. Iga, Phys. Rev. Lett. **103**, 077203 (2009)
18. R. Le Toquin, W. Paulus, A. Cousson, C. Prestipino, C. Lamberti, J. Am. Chem. Soc. **128**, 13161 (2006)
19. http://www.helmholtz-berlin.de/user/neutrons/instrumentation/laboratories/degas/index_en.html
20. V.K. Peterson, C.M. Brown, R.A. Livingston, Chem. Phys. **326**, 381 (2006)
21. V.K. Peterson, in *Studying kinetics with neutrons* ed. by G. Eckold, H. Schober, S.E. Nagler. Springer Series in Solid State Physics, vol 161 (Springer, Berlin, 2010), p. 19
22. V.K. Peterson, A.E. Whitten, J. Phys. Chem. C **113**, 2347 (2009)
23. V.K. Peterson, D.A. Neumann, R.A. Livingston, Chem. Phys. Lett. **419**, 16 (2006)
24. V.K. Peterson, M.C.G. Juenger, Phys. B **385**, 222 (2006)
25. C.G. Salzmann, P.G. Radaelli, E. Mayer, J.L. Finney, Phys. Rev. Letters **103**, 105701 (2009)
26. T.C. Hansen, M.M. Koza, P. Lindner, W.F. Kuhs, J. Phys. Condens. Matter **20**, 285105 (2008)
27. M. Koza, T. Hansen, R.P. May, H. Schober, in *Studying Kinetics with Neutrons* ed. by G. Eckold, H. Schober, S.E. Nagler. Springer Series in Solid State Physics, vol 161 (Springer, Berlin, 2010), p. 77
28. D. Ma, A.D. Stoica, X.L. Wang, Z.P. Lu, T. Proffen, Appl. Phys. A **99**, 537 (2010)
29. F. Xia, Q. Gujie, J. Brugger, A. Studer, S. Olsen, A. Pring, Rev. Sci. Instr. **81**, 105107 (2010)
30. C.R. Haramagatti, A. Islamov, H. Gibhardt, N. Gorski, A. Kuklin, G. Eckold, PCCP **8**, 994 (2006)
31. R. Lund, in *Studying Kinetics with Neutrons* ed. by G. Eckold, H. Schober, S.E. Nagler. Springer Series in Solid State Physics, vol 161 (Springer, Berlin, 2010), p. 213
32. R. Lund, L. Willner, J. Stellbrink, P. Lindner, D. Richter, Phys. Rev. Letters **96**, 068302 (2006)
33. A. Halperin, S. Alexander, Macromolecules **22**, 2403 (1989)
34. I. Grillo, E.I. Kats, A.R. Muratov, Langmuir **19**, 4573 (2003)
35. L.F. van Heijkamp, I.M. de Schepper, M. Strobl, R.H. Tromp, J.R. Heringa, W.G. Bouwman, J. Chem. Phys. A **114**, 2412 (2010)
36. G. Eckold, J. Phys. Condens. Matter **13**, 217 (2001)
37. D. Caspary, G. Eckold, F. Güthoff, W. Pyckhout-Hintzen, J. Phys. Condens. Matter **13**, 11521 (2001)
38. G. Eckold, D. Caspary, H. Gibhardt, W. Schmidt, A. Hoser, J. Phys. Condens. Matter **16**, 5945 (2004)
39. P. Elter, G. Eckold, H. Gibhardt, W. Schmidt, A. Hoser, J. Phys. Condens. Matter **17**, 6559 (2005)
40. G. Eckold, in *Studying Kinetics with Neutrons* ed. by G. Eckold, H. Schober, S.E. Nagler. Springer Series in Solid State Physics. vol 161 (Springer, Berlin, 2010), p. 178
41. R. Hill, A. Calver, A. Stamboulis, N. Bubb, J. Am. Ceram. Soc. **90**, 763 (2007)
42. J.E. Daniels, T.R. Finlayson, M. Davis, D. Damjanovic, A.J. Studer, M. Hoffmann, J.L. Jones, J. Appl. Phys. **101**, 104108 (2007)
43. G. Eckold, H. Gibhardt, D. Caspary, P. Elter, K. Elisbihani, Z. Krist. **218**, 154 (2003)
44. G. Eckold, M. Hagen, U. Steigenberger, Phase Transitions **67**, 219–245 (1998)
45. K. Elisbihani, H. Gibhardt, G. Eckold, Phys. Chem. Chem. Phys. **11**, 3168–3175 (2009)
46. V. Janovec, Phys. Lett. **99A**, 384 (1983)
47. K. Kawasaki, J. Phys. C Solid State Phys. **16**, 6911 (1983)
48. J. Leist, Thesis, University of Göttingen, 2008
49. G. Shirane, Y. Yamada, Phys. Rev. **117**, 858 (1969)
50. K.A. Müller, H. Burkard, Phys. Rev. B **19**, 3593 (1979)
51. A. Wiedenmann, U. Keiderling, K. Habicht, M. Russina, R. Gähler, Phys. Rev. Letters **97**, 057202 (2006)
52. P.K. Liaw, H. Choo, R.A. Buchanan, C.R. Hubbard, X.L. Wang, Mat. Sci. Eng. A **437**, 126 (2006)

Chapter 6
Scanning Tunneling Microscopy at Elevated Pressure

Operando Observations of Model Catalysts

Joost W. M. Frenken

Abstract Advanced experiments and theory make it increasingly clear that there can be essential differences between the behavior of catalysts under actual operation conditions and under the (near)-vacuum conditions of traditional laboratory experiments and total-energy calculations. These differences involve the structure, morphology, composition, reaction mechanisms and, importantly, the activity and selectivity of the catalyst. This forces us to cross the *'pressure gap'* with new instruments and adapted versions of existing experimental techniques. This chapter focuses on the development of dedicated scanning probe microscopy for *operando* observations of active model catalysts. The combination is advocated of an uncompromised, ultrahigh vacuum environment for the preparation and traditional characterization of model catalyst samples with a high-pressure flow reactor cell, integrated with the essential components of a Scanning Tunneling Microscope (STM) or Atomic Force Microscope (AFM). With such a configuration it becomes possible to perform atomic-scale observations on well-defined model catalysts under fully controlled working conditions. These structural observations acquire further significance by being combined with time-resolved mass spectrometry measurements on the gas mixture that flows out of the reactor, allowing one not only to measure the influence of the gas mixture on the surface structure but also to correlate that structure with the catalytic performance. This chapter highlights some of the important design considerations for high-pressure scanning probe instruments and illustrates their performance with results obtained for catalytic oxidation and reduction reactions.

J. W. M. Frenken (✉)
Kamerlingh Onnes Laboratory, Leiden University, Niels Bohrweg 2,
2333 CA Leiden, The Netherlands
e-mail: Frenken@Physics.LeidenUniv.nl

A. Ziegler et al. (eds.), *In-situ Materials Characterization*,
Springer Series in Materials Science 193, DOI: 10.1007/978-3-642-45152-2_6,
© Springer-Verlag Berlin Heidelberg 2014

6.1 Introduction

The conditions of practical catalysis, whether in industrial processes or in the three-way car catalyst, are usually at variance with the conditions that are required to perform accurate measurements with the majority of surface-sensitive techniques for microscopy and spectroscopy of the outermost atomic layers of model catalysts. Even though 'real' catalysis often involves pressures above 1 bar and temperatures above 100 °C, many fundamental studies of heterogeneous catalysis have been limited by the employed electron- and ion-based methods to pressures typically below 10^{-5} mbar and temperatures below 100 °C [1]. Although there are several cases where low-P-low-T results can be extrapolated completely up to industrial conditions [2, 3], the number of examples is growing where the 'pressure gap' of e.g. ten orders of magnitude between true process conditions and laboratory experiments is found to fundamentally change the process [4, 5].

6.1.1 Are High Pressures Necessary?

There are good reasons why there can be such important differences between low and high pressures. It is not primarily the mechanical force that is decisive, but rather the *chemical* force, or, expressed more accurately, the chemical potential of the reactants and the products to which the active surfaces are exposed. Because of the logarithmic relation between pressure and chemical potential, the influence of pressure is weak. Nevertheless, the differences in chemical potential associated with the pressure gap, i.e. with pressure changes of 10 orders of magnitude or more, are significant and can easily be sufficient to stabilize new surface structures and surface compositions [6–9]. In other words, a surface configuration that has a high formation energy and is therefore unfavorable at low pressures, may become the preferred phase, i.e. the structure with the lower *free* energy, at high pressures. In such cases, a first-order phase transition will occur in the surface configuration at a certain pressure. If more than a single surface composition can be formed, there can be several such phase transitions, each at a different pressure (chemical potential). In most catalytic reactions, the catalyst is exposed to a mixture of two (or more) reactants. Even if we can neglect the presence of the reaction products, we need to consider the two-dimensional phase diagram of the surface as a function of the partial pressures of the reactants, taking into account possible co-adsorption structures and ternary (or even more complex) surface compounds. Figure 6.1 illustrates this scenario with a phase diagram calculated by Rogal et al. [10, 11] for one of the model systems that will be discussed in this chapter.

What is clear from the above description is that the difference between low and high pressures (chemical potentials) is not a trivial matter of longer or shorter exposure times to the reactants; whether or not a structure is stable does not depend on the patience of the experimentalist. In an experiment at low pressures,

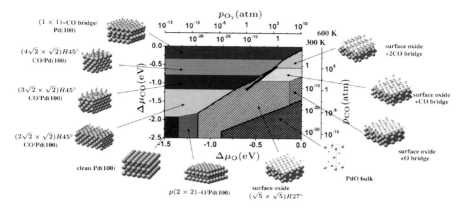

Fig. 6.1 Phase diagram calculated by a combination of Density Functional Theory and thermodynamics for the Pd(100) surface in contact with mixtures of CO and O_2. At low pressures, the clean metal surface is stable. When only CO is introduced, a variety of adsorption structures is obtained, depending on the chemical potential (*left vertical axis*). The axis on the *right* translates the CO chemical potential into the partial pressure of CO for two different temperatures. When the surface is exposed to increasing amounts of O_2 (*horizontal axis*), it progresses through a sequence with an adsorption layer, a thin surface oxide and, finally, the stabilization of the bulk oxide, PdO. When exposed to mixtures of CO and O_2, the surface adopts various additional CO adsorption structures on the surface oxide. The *thick black line* indicates the experimental regime of catalytic conditions for this system. (From [10, 11]; figure courtesy of Prof. Karsten Reuter, Technische Universität München. Copyright 2007 by the American Physical Society)

the model catalyst may simply be in the 'wrong' phase. This is why it is necessary to traverse the pressure gap and perform our experiments under the relevant, high-pressure conditions.

6.1.2 Why Can't We Use the Thermodynamic Back Door of Temperature?

Thermodynamics provides a second way to make the chemical potential high, namely by lowering the temperature. This is illustrated implicitly by the two temperature scales in Fig. 6.1. This implies that the high-pressure (high-temperature) structures should also be stable at low pressure when the temperature is made low enough. However, this approach comes with several risks. First of all, the activation energy for the restructuring into the most favorable configuration may be too high for a transition to take place on a reasonable time scale at low temperatures, as may be expected in cases that require a significant sub-surface rearrangement of the model catalyst. Secondly, the catalytic reaction itself may be important in keeping the system permanently out of the equilibrium of the phase diagram (Fig. 6.1). If the reaction runs by virtue of the deviation from equilibrium, e.g. the disorder introduced by the reaction itself, as has been suggested by the computer simulations by Rogal et al. [12], the exponential reduction of the reaction rate that accompanies the lowering of the temperature will have a dramatic effect that is not captured by the value of the chemical potential. Finally, each catalytic process involves several thermally activated steps, such as dissociation, diffusion, reaction, and desorption, each step having its own activation energy. When the temperature is lowered the rate at which each of these steps proceeds is reduced by a different factor. This means that the overall reaction will not just run more slowly (or even effectively come to a stand-still), but it will do so in a manner that is not representative for the situation during the active process at high temperatures and high pressures.

What we learn from these arguments is that there is no general guarantee for the existence of a niche of low-temperature-low-pressure conditions where we can observe the same processes in 'slow motion' that make a catalyst work under practical conditions. There is no 'escape': we are simply forced to modify our measurement techniques so that we can use them under realistic reaction conditions.

6.2 High-Pressure Techniques

The last decade has witnessed a rapid growth of the number of surface-science techniques that are adapted to bridge the pressure gap in some way or another. A natural approach is to combine a reaction zone at high pressures with traditional ultrahigh-vacuum (UHV) instrumentation by use of differential pumping. One can extend the pressure ranges of techniques such as X-Ray Photoelectron Spectroscopy (XPS) [13, 14], Low-Energy Ion Scattering (LEIS) [15], and Transmission Electron Microscopy (TEM) [16, 17], into the millibar regime. A recent review of several of these developments can be found in [18]. A new development is that of micro- and nanoreactors that separate a small high-pressure volume from a low- or ultrahigh vacuum environment by ultrathin walls of an inert material, such as SiN. This approach has been demonstrated in TEM [19] and X-Ray [20] microscopy experiments.

A few experimental methods are relatively robust with respect to the presence of a gas atmosphere and can be applied to catalytic surfaces at atmospheric pressures. Examples are surface-sensitive optical spectroscopy and various microscopy techniques [21–26]. X-Ray Absorption Spectroscopy [27, 28] and Surface X-Ray Diffraction (SXRD) [29, 30] have been used successfully to resolve the electronic and atomic structures of catalytic surfaces at elevated pressures in single-component gasses or during reaction.

For most forms of scanning probe microscopy, in particular for Scanning Tunneling Microscopy (STM) and Atomic Force Microscopy (AFM), there are no fundamental reasons why they cannot be employed with high resolution for the investigation of model catalyst surfaces, even under the harsh conditions of industrial catalysis. The distance between the apex of the tip of the scanning probe instrument and the catalyst surface is so small that on average there will be less than one gas phase molecule between the two that could significantly alter the probe signal, i.e. the tunneling current or the interaction force. However, there are several practical aspects that can seriously affect the imaging quality and resolution.

The following sections review the design and performance of the so-called 'ReactorSTM', a dedicated instrument that combines a miniature flow reactor with a compact scanning tunneling microscope [31–33]. Most of the examples presented here have been obtained with the 'first-generation' version of this instrument. Recently a new, improved ReactorSTM has been completed in our laboratory. It is the design of the latter instrument that will be discussed and a few recent results obtained with it are provided to illustrate its performance. The two ReactorSTMs have provided us with a first direct, atomic-scale look on surfaces under combined high-pressure, high-temperature conditions. This has led to surprising, new insight in the thermodynamics and kinetics of surfaces under such conditions.

6.3 High-Pressure STM Instrumentation

The most straightforward approach towards STM observations at atmospheric pressures of reactive gasses is to use a microscope in an ultrahigh vacuum (UHV) chamber that is backfilled with the gas after preparation of the sample. This has been demonstrated first by McIntyre et al. for the Pt(110) surface in atmospheric pressures of H_2, CO and O_2 [34] and it has been adopted by several other research groups [35–39], to image adsorbate structures and gas-phase induced restructuring of metal surfaces and supported metal nanoparticles [40–54]. In order to routinely operate a high-pressure STM also at elevated temperatures in a reactive gas mixture, i.e. *during* a catalytic process, several conditions need to be met that are briefly summarized below.

6.3.1 Requirements for STM Under Catalytic Conditions

For the ReactorSTM that we have developed for dedicated STM observations *during* catalytic reactions, we have adopted a rather unconventional design. The configuration of this instrument stems from the combination of the following requirements.

(1) The STM should operate under reaction conditions, i.e. while chemical reactions are taking place on the model catalyst surface at elevated temperatures (100 °C or higher) and atmospheric pressures (1 bar or higher). Under these conditions, the microscope should be capable of imaging the active surface with atomic resolution and with acceptably low drift, low enough to enable one to track individual features on the surface.

(2) The instrument should enable the combination of the imaging of the model catalyst surface with simultaneous, accurate measurements of the catalytic performance, so that the two can be correlated with a meaningful time resolution (10 s or better), allowing one to follow the chemical effects accompanying variations in surface structure.
(3) The configuration of the instrument should be chosen such that it minimizes undesired chemistry in the form of chemical reactions taking place elsewhere than on the model catalyst surface; this might either contaminate the gas mixture that is to be analyzed in order to characterize the catalytic performance or even corrode components of the microscope.
(4) Measurements should be possible on well-prepared model catalyst surfaces that are at least initially highly ordered and atomically clean. Between preparation and experiment the model catalysts should not be exposed to air.

Before introducing the actual design of the ReactorSTM, let us briefly consider these requirements and their implications.

6.3.1.1 Atmospheric Pressures

As remarked before, the scanning tunneling microscope has no intrinsic difficulty with imaging surfaces at atmospheric and higher pressures. Since the technique even works routinely for imaging surfaces in liquids, we can be confident that the mere presence of a gas will not interfere unacceptably with the tunneling effect and disable imaging with atomic resolution. One point of concern, however, is the degree to which the gasses modify the apex of the STM tip, because this may directly affect the imaging quality. To our knowledge, a systematic search for optimal tips for imaging under high-pressure and high-temperature conditions has not been conducted yet.

6.3.1.2 Elevated Temperatures

Even though temperature changes lead to expansions or contractions of the mechanical components of most STM setups and therefore form the potential source of strong drifting of the images, a wide variety of STMs exist that can operate at either low or high temperatures [55, 56]. Usually this is achieved by the use of special materials with low expansion coefficients, symmetric STM geometries, and by allowing the instrument to settle after each change in temperature, before the tip engages the surface. For the development of a dedicated variable-temperature STM, our group has undertaken a full optimization of the thermal behavior of the microscope and specimen by a complete, finite-element analysis, which has resulted in an instrument that can routinely image surfaces during a full temperature sweep from room temperature to 1300 K and higher [57].

6.3.1.3 Atmospheric Pressures Combined with Elevated Temperatures

Whereas high pressures and high temperatures may not pose too much of a problem for STMs separately, their combination is challenging. One problem is that the piezo elements that are used in most STMs to actuate the tip must be kept below their Curie temperature, also when the sample is much hotter. This is no problem in vacuum, where temperature differences of hundreds of degrees can easily be maintained over short distances, but the heat transport through the gas atmosphere makes it difficult to do the same at high pressures. When nevertheless a strong temperature difference is maintained between the hot specimen and the nearby piezo element, the gas will be set into a convective flow. We have observed that the accompanying thermal transport is not constant in time. It fluctuates, which we ascribe to variations in time of the convection pattern. As a consequence, the temperatures of the specimen, the tip, the piezo element and the other components that are part of the mechanical loop between sample and tip are all subject to small variations with characteristic times in the order of a few seconds. The consequences are recognizable in the STM images as erratic drift, changing direction and strength several times within a single image. One approach to avoid this problem is to keep most of the components of the STM in vacuum, while the specimen is exposed to reaction conditions.

6.3.1.4 Correlation with Catalytic Performance

The requirement to measure the chemical activity simultaneously with the imaging of the model catalyst surface makes it necessary to treat the environment of the catalytically active surface as a reactor. Here, we distinguish two conceptually different reactor types. A flow reactor operates with a steady supply of reactants and it has an exhaust through which products and unused reactants continuously leave the reactor. A batch reactor is a closed system that starts with a mixture of reactants that is consumed in the reaction, while the reaction products remain stored in the same container. The catalytic activity can be determined by analyzing the composition of a small fraction of the gas that is allowed to leak from the exhaust (flow reactor) or from the reactor itself (batch reactor) into a mass spectrometer. In order to optimize the combination of the time resolution in this mass spectrometry and the sensitivity to changes in the concentrations of reactants and products, we need to maximize the ratio between the surface area of the model catalyst and the volume of the reactor. Here, we are immediately confronted with an intrinsic drawback of scanning probe microscopes, which is that they cannot be used on the complex, three-dimensional architectures of most practical catalysts. Instead, we are limited to nearly perfectly flat surfaces or ensembles of nanoparticles on flat supports and therefore we are limited also to an extremely small area of catalytically active surface. The only way for us to make the active-surface-area-to-reactor-volume ratio high is to make the reactor volume as small as possible.

6.3.1.5 Minimal Chemical Side Effects

Under the high-pressure, high-temperature conditions of the catalytic processes, chemical reactions can take place not only the surface of interest of the model catalyst but also on the side and rear faces of the catalytic specimen and on other components that are exposed, such as the specimen holder, the heating element and the essential scanning probe microscope components, such as the electrical wiring, the piezo electrodes, the tip holder and the tip. Obviously, such parasitic chemistry can damage the instrument. It can also lead to contributions to the mass spectrometry signals that stem from the instrument and from the differently oriented sides of the specimen, rather than the model catalyst surface that is imaged by the microscope. These aspects favor a design in which only the catalytic surface of interest is brought in contact with the gasses, while the rest of the specimen and most of the microscope is not exposed.

6.3.1.6 Well-Prepared Model Catalysts

Standard methods to clean a model catalyst surface, such as ion sputtering, annealing, metal deposition, and gas exposure, require high vacuum. Techniques to check the crystal structure and quality, such as low-energy electron diffraction, and the surface composition and cleanliness, such as Auger electron spectroscopy, require high or ultrahigh vacuum. Between preparation and catalytic experiments, the specimens should not be exposed to air. This requirement calls for a combination of an ultrahigh vacuum system with the high-pressure reactor and the STM.

6.3.2 Design of the ReactorSTM

The combination of the above design criteria is satisfied by the special geometry of the ReactorSTM [5, 31–33], which is illustrated schematically in Fig. 6.2. The heart of the setup is formed by the integration of a tiny flow reactor with a scanning tunneling microscope. This combination is placed inside an ultrahigh vacuum system, in which model catalyst specimens can be prepared, characterized by conventional surface-science techniques and transported without breaking the vacuum. During STM experiments, the reactor volume is sealed off by the specimen itself, which acts as one of the reactor walls and is pressed against the reactor body via a Kalrez seal. The specimen is held in a holder that further carries a heating element behind the specimen and a reliable temperature measurement, directly attached to the specimen. During experiments, the specimen forms the hottest component of the reactor. Two generations of the ReactorSTM have been developed. In the first, that we will refer to as Mark I, the reactor body was made out of a low-expansion metal (Invar), while in the second, Mark II [33], we have replaced the metal by low-expansion glass (Zerodur). The metal version was

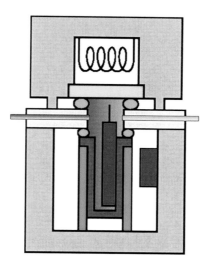

Fig. 6.2 Schematic cross section of the central part of the 'ReactorSTM'. The instrument can image a surface, while it is active as a catalyst, under gas flow conditions at pressures up to 5 bar and temperatures up to 600 K. Apart from the surface of the sample, only the tip of the STM is in contact with the flowing, hot, high-pressure gas mixture. The specimen (*yellow*) serves as the *upper wall* of the small reactor volume. Together with the filament that heats it from the rear and the thermocouple that is used to monitor its temperature, the specimen is held in a holder (*upper grey block*) that can be manipulated through the surrounding ultrahigh vacuum system for sample exchange, preparation and characterization. The tube shaped piezo element (*green*) is used both to actuate the xyz-scan motion and to drive the coarse approach of the STM tip towards the surface of the specimen. For that purpose, the tip holder (*dark blue*) can be made to perform up- and downward stick–slip motion along two rails inside the double shield (*dark grey*) that is attached to the top of the piezo element. The tip holder is gently pulled against these rails by a small magnet (*purple*). Two O-rings (*light blue*) seal the reactor from the surrounding ultrahigh vacuum. The flexibility of the lower seal is sufficient not to frustrate the scan and coarse approach motion of the piezo element. Two capillaries (*red and light blue*) serve as the gas supply and exhaust channels. The total volume of the cell is as small as 0.5 ml

coated with a non-reactive metal film (gold), while the glass is non-reactive by itself for the oxidation–reduction reaction systems that we have investigated so far. The only components of the STM that are inside the reactor are the STM tip and its holder. In this way we have been able to limit not only the volume of the reactor to a minimum (approximately 500 μl), but also the number of materials that are exposed to high pressures and high temperatures. The piezo element that actuates the tip motion is pressed against the reactor from the outside via a small Viton O-ring that closes the reactor off from below but is flexible enough to allow the piezo element and the connected tip holder to move. During catalytic experiments, only the inside of the reactor is brought up to atmospheric pressures, while ultrahigh vacuum is easily maintained on the outside with the use of an ion pump. The ReactorSTM combination is mounted inside the ultrahigh vacuum system on a single conflat support flange in a conventional spring suspension configuration, combined with eddy current damping. Two capillaries, one for the gas supply and one for the exhaust gasses, connect the flow reactor cell to connections on the support flange. These capillaries are sufficiently thin and flexible not to strongly couple external vibrations into the STM.

We have developed a dedicated gas cabinet, with pressure regulators, flow controllers and tailor-made mixing valves, to prepare mixtures of up to 4 different, high-purity gasses, over wide ranges of partial-pressure ratios. The gas exhaust of the ReactorSTM is connected to a separate pumping stage. By controlling the pressures on the in- and outgoing gas lines separately, we can independently set the operating pressure in the reactor and the gas flow rate, i.e. the residence time of the gas in the reactor. Special care has been given to minimize the volume of all gas lines and to avoid dead volumes in the gas system. Under typical flow conditions, the time delay between changes in gas composition in the gas cabinet and in the reactor is as short as a few seconds. A small fraction of the gas leaving the reactor via the exhaust gas line is dosed into a separate ultrahigh vacuum chamber with a quadrupole mass spectrometer. With this we can accurately determine the composition of the exhaust gas, again with a short time delay of typically a few seconds.

The combination of materials, the mechanical construction and the temperature distribution in the setup are such that experiments can be performed at specimen temperatures up to 600 K (limited by the Kalrez seal) and pressures up to at least 5 bar (currently limited by the gas system; see below). The low-expansion materials and the compact, symmetric configuration result in low thermal drift, even during modest variations in temperature. Since there is no gas atmosphere outside the reactor, the piezo element does not exhibit erratic drifting due to convection-induced temperature variations. The mechanical rigidity of the microscope-reactor assembly in combination with the spring suspension system makes the STM fit for imaging with atomic resolution on a close-packed metal surface. With this we fully satisfy the first requirement of uncompromised operation under reaction conditions. As a result of the small reactor volume, we also meet the second requirement: even for relatively slow reactions, e.g. 1 conversion per site per second, the converted fraction of the gas (at atmospheric pressure) is sufficient to be detected by a QMS, also when the gas resides in the reactor only for a few seconds. The reactor geometry is also such that most of the STM components are not in contact with the high-pressure gas mixture. Also the specimen has only a single surface exposed to the gas. Together this makes the design also optimal for fulfilling the third requirement: no undesired chemistry. Finally, the setup with the microscope and specimen inside an ultrahigh vacuum system enables us to transfer the specimen in-situ between the ReactorSTM and various other techniques inside the system, for preparation (ion sputtering, annealing, metal deposition, gas exposure) and characterization (low-energy electron diffraction, Auger electron spectroscopy, X-ray photoelectron spectroscopy), thus satisfying the fourth requirement of enabling catalytic STM experiments on well-prepared and well-characterized model catalysts.

Figures 6.3, 6.4 and 6.5 are photographs of the ReactorSTM setup. The complete UHV system, including three separate chambers for analysis, preparation and high-pressure STM experiments, is shown in Fig. 6.3, together with the gas cabinet, the electronics and the control system. The central STM part, including

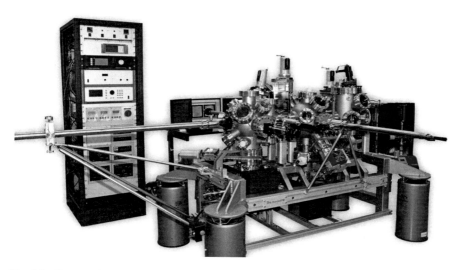

Fig. 6.3 Commercial version of the complete ReactorSTM setup. The vacuum setup consists of three separate UHV chambers for STM imaging (*right*), preparation (*center*) and characterization (*left*), connected via a linear transfer arm. The entire system is supported on air legs to reduce vibrations. Also shown are the electronics (*left*) and the gas cabinet (*blue unit on the right*) with which a wide range of gas compositions, pressures and flows can be set. (Photo courtesy of Dr. Gertjan van Baarle, Leiden Probe Microscopy B.V. [58])

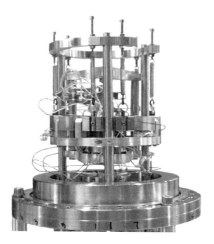

Fig. 6.4 Flange with the completely assembled, central part of the ReactorSTM. The copper block that serves as the mounting plate for the reactor body with the STM components is suspended by six springs from the upper ring and its motion is reduced by eddy current damping. In addition to electrical connections for the heating, temperature measurement and STM control, capillaries run from the bottom flange to the reactor body as gas supply and exhaust lines. (After [33])

Fig. 6.5 View on the ReactorSTM assembly with the reactor opened. The sample holder is the part that is colored *green* in the inset. The reactor body (*red part*) has been lowered to open the reactor. The Kalrez seal is attached to a metal strip (together forming the *yellow part*) that enables one to pick it up and exchange it easily. The ReactorSTM is closed in a smooth way by inflating a bellow that is located below the reactor body. (After [33])

all electrical connections and gas capillaries, is mounted on a single flange. This part of the setup is shown in Fig. 6.4. What can be recognized is the spring suspension combined with the copper of the eddy current damping unit, with which the mechanical coupling between the microscope and bottom flange is reduced. The gas capillaries run to the in- and outlets of the reactor body in a gentle loop, thus minimizing their contribution to the mechanical coupling. The construction in Fig. 6.4 further contains a mechanism that is used to push the reactor body upwards against the sample holder, in order to make mechanical contact between the reactor and the sample and to simultaneously seal off the reactor volume from the UHV chamber. An additional mechanism is used to lock the suspension system during the introduction and removal of either sample holders or Kalrez seals.

The photograph in Fig. 6.5 shows the sample holder (colored green in the inset), the Kalrez seal (yellow) and a view on top of the reactor body (red). The sample holder can be made of various low-expansion-coefficient materials. The one shown here was made of invar. The sample holder contains a double set of electrical connections, for heating, temperature measurement and STM bias voltage. The reactor body was machined from Zerodur glass. The actual reactor is a cylindrical hole, directly below the Kalrez ring in the center of the body; the tip holder and tip of the STM are inside. One of the gas capillaries can be recognized in the front. The ring-shaped Kalrez seal was vulcanized onto a metal strip that allows us to exchange such seals with a wobble stick, without the necessity to break the vacuum of the STM chamber.

6.4 High-Pressure Observations

In this section we illustrate the observations that are possible with the ReactorSTM for two catalytic reactions, namely the oxidation of CO and the reduction of NO. Most STM images and movies obtained thus far, have been acquired with the Mark I version of the ReactorSTM. In most cases, the resolution in these images is just insufficient to distinguish individual molecules, atoms, or atom rows, but it is good enough to recognize the terraces and steps, which already provides a wealth of useful information. One example will be shown below of the images obtained nowadays with the Mark II instrument, in which much more detail of the surface structure is resolved.

6.4.1 CO Oxidation: The Special Role of Surface Oxides

In Figs. 6.6, 6.7 and 6.8 results are shown that have been obtained on the Pt(110) surface during the oxidation of CO by O_2, one of the reactions running in the three-way car catalyst. Figure 6.6 contains a sequence of images, acquired with the platinum surface at an elevated temperature of 425 K. The sequence starts after the surface has been brought into contact with a high pressure of 1.25 bar CO. Whereas the clean surface has been shown to exhibit the missing-row reconstruction, with every second Pt row in the surface missing [59], low pressures of CO make the surface adopt an unreconstructed geometry [60]. Under the high-pressure, high-temperature conditions of Fig. 6.6, the surface also prefers the unreconstructed geometry, which initially brings the surface into the problematic situation of being only half-occupied. As a consequence, the CO-exposure immediately introduces a high density of steps on the surface, as the surface is broken up into two height levels everywhere, corresponding to the present and the absent rows in the initial missing-row structure. Edge diffusion of Pt atoms leads to a coarsening of this step structure and eventually to the complete removal of these extra steps [32, 61].

In this particular case, the structure of the metal surface at high pressure is not different from that at low-pressure exposure to the same gas, even though the density of the CO overlayer may depend on the CO pressure [63]. The higher-resolution images in Fig. 6.7 show the close-packed atom rows of the missing-row reconstruction of the initial surface and the doubled density of atom rows of the CO-induced unreconstructed surface, as obtained with the Mark II version of the ReactorSTM.

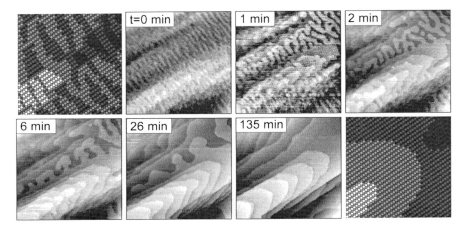

Fig. 6.6 Series of STM snapshots (140 × 140 nm) taken on Pt(110), starting immediately after introduction of 1.25 bar CO in the Mark I ReactorSTM at 425 K. The 'tiger skin' pattern in the *first panel* shows that the CO-induced removal of the 'missing-row' reconstruction has divided the surface in two levels, each 50 %, and a high density of steps. Subsequent images show the progressive reduction of the step density by coarsening of the step pattern. The elapsed time in minutes is indicated in each panel. The two ball models indicate the atomic-scale geometries characteristic for the starting and end situations. (After [61]. Copyright 2005 by Springer Verlag)

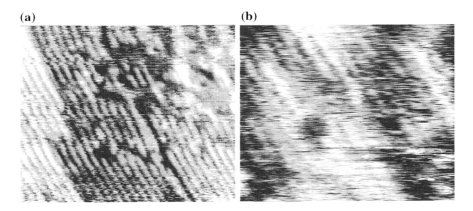

Fig. 6.7 High-resolution images obtained with the Mark II ReactorSTM on Pt(110). **a** The (1 × 2) missing row reconstruction of the clean surface observed a short time after placing the platinum specimen on the reactor (22 × 19 nm). The high density of defects in the structure results from the adsorption of contaminants, such as CO in the poor vacuum of the reactor. **b** The (1 × 1) unreconstructed surface in 1 bar of CO at 160 °C (5.9 × 5.0 nm). The imaged region contains many height defects that result from the (1×2) to (1×1) transition and that are slowly removed with time, as is seen on a larger length scale in Fig. 6.6. Note that the scales of the two images are rather different. (After [33])

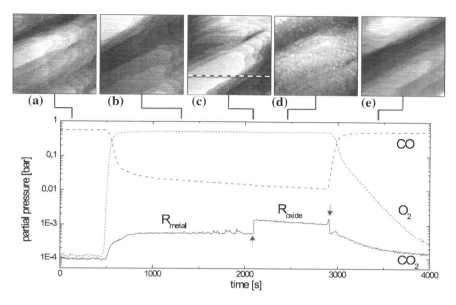

Fig. 6.8 Images obtained with the Mark I ReactorSTM and mass spectrometer signals measured simultaneously during CO oxidation on Pt(110) at a temperature of 425 K in a 3.0 ml/min flow of a CO and O_2 gas mixture at a total pressure of 0.5 bar. (*Upper panel*) A selection of STM images (210 × 210 nm) from an STM movie [62]. Images **a**, **b**, the lower part of **c** and image **e** show the metallic, mainly CO-covered Pt(110) surface. The *upper part* of image **c** and image **d**, represent the oxide surface which gradually roughened. (*Lower panel*) Partial pressures of the reactant gasses CO and O_2 and the reaction product CO_2. R_{metal} denotes the reaction rate on the metallic surface and R_{oxide} is the rate on the oxide surface. The arrows indicate where the surface switched from metal to oxide and back. (Figure adapted from [5]. Copyright 2002 by the American Physical Society)

The powerful combination of simultaneous imaging and mass spectrometry under reaction conditions is illustrated in Fig. 6.8, which shows the (110) surface of Pt in contact with several CO/O_2 mixtures at atmospheric pressures and elevated temperatures. In addition to the unreconstructed structure of Figs. 6.6 and 6.7, characteristic for a CO-covered, metallic surface, which we find in CO-dominated gas mixtures, the images reveal the presence of an atomically thin surface oxide in strongly O_2-dominated mixtures. STM-images obtained during slow variations in the gas composition show that the transitions between the two structures, from metal to oxide and vice versa, are all very abrupt. They happen within a single scan line, well within the STM time resolution of ~1 s, thus supporting the 1st-order character of the transition, expected on the basis of the thermodynamic considerations discussed in Sect. 6.2.

Figure 6.8 also displays the calibrated mass spectrometer signals that were collected simultaneously with the images. The signals for the three most important masses are shown here, corresponding to the reactants, CO and O_2, and the reaction product, CO_2. When we vary the partial pressures of CO and O_2, the partial pressure of CO_2, which is directly proportional to the CO_2 formation rate, responds by going up or down in a continuous fashion. However, precisely at the points in time where the STM images show that the surface switches from metal to oxide or from oxide to metal, the CO_2 signal exhibits an abrupt change. The reaction rate that we observe when the surface oxide is present is higher than that on the metal surface.

High-resolution images, obtained with the Mark II ReactorSTM under reaction conditions, with the Pt(110) surface in the oxide phase, reveal a regular (1 × 2) pattern [33]. This pattern has also been observed in separate surface X-ray diffraction experiments on this surface under the same conditions [64]. The crystal truncation rods, measured in this experiment indicate that every second row of Pt atoms is pushed upwards and shifted sideways over substantial distances with respect to its position for the unreconstructed surface. It is probably these rows that stand out in the high-resolution STM images. Density functional theory calculations by Hammer and coworkers explain this unconventional structure by the presence of a row of carbonate ions (CO_3^{-2}) below the row of high Pt atoms plus a separate row of oxygen atoms adsorbed on top of the surface [64]. Whether the name "surface oxide" is appropriate for this special surface layer should be regarded as a semantic issue.

The scenario of Fig. 6.9 explains these STM observations of the Pt(110) model catalyst during CO oxidation at atmospheric pressures and elevated temperatures. When the surface is in the metallic state, the CO oxidation reaction proceeds according to the Langmuir–Hinshelwood (LH) mechanism [65], familiar from detailed investigations of many catalytic reactions at low pressures (high vacuum). In the LH reaction, CO and O_2 compete for adsorption sites on the surface. If a CO molecule and an O atom are adsorbed on neighboring sites, either by direct adsorption or by diffusion from other adsorption positions, they can react with each other to form CO_2 that will desorb from the surface after some time. When the surface is made to switch from metal to oxide, the reaction switches from the LH mechanism to an alternative mechanism that can be regarded as the surface analog of the well-known Mars-van-Krevelen (MvK) mechanism [66]. In the traditional MvK mechanism an oxidation reaction makes use of the oxygen atoms that are part of an oxide matrix. The resulting oxygen vacancies in the oxide (the 'missing' oxygen atoms) are refilled by oxygen that is supplied from the gas phase, the newly arriving oxygen atoms diffusing through the oxide to annihilate these vacancies. In the present case, the latter diffusion step is not necessary, as the oxide is atomically thin; the oxygen vacancies are formed and removed at the very surface, where there is direct contact both between the oxygen atoms and the CO and between the oxygen vacancies and the O_2. The kinetics of the two reaction mechanisms, LH and MvK are quite different. In the regular LH situation, the rate of CO_2 formation depends nonlinearly on the partial pressures of both reactants, CO and O_2, while in the MvK reaction, the rate depends linearly on the partial pressure of CO and is independent of the O_2 pressure: after all, the relevant oxygen content corresponds to the fixed density of O atoms in the oxide rather than the density of O_2 molecules in the gas phase. The rates observed on Pt(110) and on other surfaces that display the same behavior are fully consistent with these kinetics.

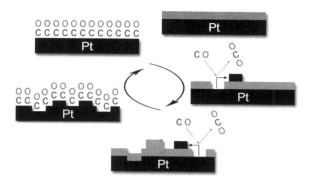

Fig. 6.9 Schematic of reaction mechanisms for CO oxidation on Pt(110). At low partial pressure of oxygen the Pt(110) surface is smooth and almost fully covered by a dense, CO overlayer (*upper left*). The CO oxidation reaction proceeds via the Langmuir–Hinshelwood mechanism with dissociatively adsorbed oxygen. At sufficiently high oxygen pressure a thin surface oxide is formed (*upper right*). Then, the reaction takes place between CO molecules and oxygen atoms that are removed from this oxide (*center right*). Occasionally, an under-coordinated Pt atom diffuses out of the reduced region, to be oxidized on top of the oxide, making the oxide surface increasingly rough (*bottom*). When exposed to a high CO pressure, the rough oxide is removed by the reaction; a rough metallic surface remains (*center left*), which quickly becomes smooth to restore the initial situation (*upper left*). (After [61])

In addition to the measurements described here on the (110) surface of Pt, we have investigated the oxidation of CO on Pt(111), Pt(100), Pd(100) and on surfaces vicinal to Pd(111) and Pd(100) [67–69], at atmospheric pressures and elevated temperatures. In all cases we have encountered similar scenarios, with the surface switching upon increasing partial pressure of O_2 or decreasing partial pressure of CO from a metallic state, covered by adsorbed CO molecules and/or O atoms, to a surface oxide or thin bulk oxide, the reaction mechanism switching accordingly from LH to MvK. For several of these cases, we have performed experiments in parallel with *operando* SXRD, supporting these conclusions and in each case providing detailed information on the thickness and atomic structure of the oxides [64, 70, 71]. The formation of a surface oxide or thin bulk oxide on these metal surfaces should be regarded as a genuine *pressure-gap* effect. With the exception of the surface oxide on Pd(100), these surface oxide structures are not observed under the low-pressure conditions of typically 10^{-5} mbar and lower of most surface-science experiments, even when the low pressures are combined with a matching low temperature (see Sect. 2.2).

6.4.2 CO Oxidation: Roughness, Steps and Spontaneous Reaction Oscillations

A salient feature of the STM images during the CO oxidation reaction in the oxide phase on all metal surfaces mentioned above is that they show the steady buildup of a significant surface roughness, which was shown in the case of Pt(110) to scale more or less linearly with the total amount of CO_2 produced [32, 61]. Such roughening is observed only under reaction conditions, when the surface is in contact with both CO and O_2, at atmospheric pressures and elevated temperatures. This kinetic roughening has also been observed with SXRD measurements [64] and can be explained naturally in the context of the MvK mechanism. STM movies [62] in a pure O_2 atmosphere show that at the temperatures of these experiments, the oxides on the Pt and Pd surfaces exhibit very low mobilities, whereas the metallic surfaces in a pure CO or CO-dominated atmosphere are rather mobile. This implies that every mechanism that leads to height variations on the oxide will result in the accumulation of roughness. The MvK mechanism itself, i.e. the removal of O atoms by CO molecules, cannot be responsible for the roughening. Every time that an oxygen vacancy is filled up, the flat surface oxide is locally restored again. The STM movies and SXRD measurements have shown that the roughness involves also the metal atoms. These observations all strongly suggest that the metal atoms near a (single or multiple) oxygen vacancy have a small, but non-zero probability for diffusing out of the oxide matrix. In the high partial pressure of O_2, such a diffusing metal atom will be oxidized, and thus immobilized, very quickly. In this way, a new, somewhat rougher surface contour is 'frozen' in, every time that a metal atom is relocated. In the case of Pt(110), we have estimated that on average one such a relocation of a Pt atom takes place per 500 CO_2 molecules formed.

The combination of the kinetic roughening that accompanies the MvK mechanism on the surface oxide with the smoothening that results from the rapid surface diffusion in the metal phase (see Fig. 6.6) can lead to an interesting instability that makes the surface structure and the reaction rate oscillate spontaneously [61]. An example of such a spontaneous oscillation is visible in Fig. 6.10 in the variations of the partial pressures of the reaction product CO_2 and the reactant CO during catalytic oxidation of CO on Pd(100) [72].

The character of these oscillations is that of a sequence of abrupt transitions between two distinct reaction rates, the lower corresponding to the LH rate and the higher to the MvK rate that were discussed above [61, 72]. The STM images recorded during these oscillations indeed indicate that the surface is switching back and forth between the corresponding structures: metal and surface oxide. The same, first-order transitions between metal and surface oxide and vice versa were seen in our surface X-ray diffraction experiments during reaction oscillations on Pd(100) under the same conditions [72]. At first sight it may seem difficult to reconcile spontaneous oscillations with the two-phase character of the first-order transitions. Any mechanism that would lead to such a periodic sequence of first-order transitions necessarily features at least one parameter that evolves and oscillates continuously during the process. It should be this 'hidden' parameter that sets the characteristic time between the first-order transitions and thus determines the oscillation period. The only property that changes continuously in our STM images is the surface roughness; as explained above, it increases continuously during the MvK phase of the oscillation cycle, whereas it decreases continuously during the LH phase. We explain this via the thermodynamics of the first-order transition

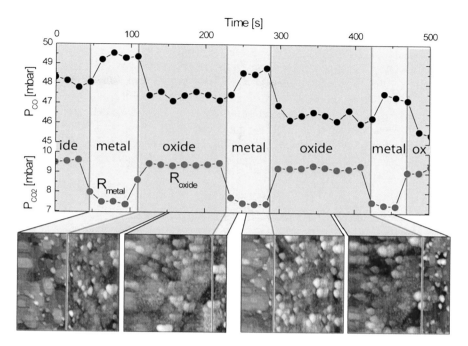

Fig. 6.10 Spontaneous oscillations in the CO oxidation rate on Pd(001). The *top panel* shows the reaction rate (P_{CO2}) and the CO pressure (P_{CO}), observed in the Mark I ReactorSTM at a constant oxygen pressure of $P_{O2} = 1.2$ bar and a temperature of 408 K. The STM images (140×140 nm, slow scan direction along the time axis) have been recorded simultaneously with the CO_2 and CO pressures and show that the oscillations are accompanied by the oxidation and reduction of the Pd surface. The metal phase exhibits characteristic terrace-and-step configurations with the well-defined step height of Pd(001), while the oxide is rougher and shows no such order. (Figure from [72]. Copyright 2010 by Macmillan Publishers Limited)

between the metal surface and the surface oxide. The observations suggest a direct influence of the surface roughness on this transition. Additional surface X-ray diffraction measurements, in which the Pd(100) surface was forced through the transition at different values of the surface roughness, which can be quantified via the width of a surface diffraction profile, show that this is indeed the case: a rough surface oxide requires a lower partial pressure of CO to switch to a (rough) metal surface than a smooth surface oxide. The rational behind this relation is that CO molecules bind more strongly to the steps on the metallic Pd(100) surface than to terrace sites, whereas such a preference for steps is not present or much weaker on the surface oxide. As a consequence, the free energy difference between the metal and oxide surfaces, in contact with the CO containing gas mixture, is a direct function of the step density. This leads to a window of partial pressure ratios between CO and O_2 in which the smooth metal surface is not stable, but the surface oxide can develop sufficient roughness to make it switch spontaneously to the rough metal surface, which is necessarily smoothening due to surface diffusion. In this scenario, the typical time that the surface spends in the oxide phase is determined by the precise setting of the partial pressures of the reactants and by the rate at which the oxide roughens, which is a super-linear function of the reaction rate and a weak function of temperature (because of the modest mobility of the oxide). The average time that the surface spends in the metal phase is determined mostly by the step removal rate, i.e. by the edge diffusion rate of the CO decorated step atoms, which depends strongly on temperature an only very weakly on the partial pressures of the reactants. These simple expectations are borne out by our experiments. A crude, numerical simulation of the oscillation mechanism described here, provides a nearly quantitative fit of our experimental observations [72].

6.4.3 NO Reduction: From LH to LH

Also the reduction of NO by CO is one of the reactions in the three-way car catalyst, running simultaneously with the oxidation of CO and of hydrocarbons. We have used the Mark I ReactorSTM to image the Pt(100) surface during this reaction at atmospheric pressures and at elevated temperatures, under various ratios between the partial pressures of the reactants, NO and CO [73]. Some of the results from this investigation are given in Figs. 6.11, 6.12 and 6.13. Like Pt(110), the clean (100) surface of platinum exhibits a reconstruction. The outermost layer adopts a quasi-hexagonal symmetry (*hex*-reconstruction), thus allowing itself to acquire an areal density of Pt atoms that is 20 % above that of a non-reconstructed (100) plane of the platinum lattice. It is known that low pressures of either NO or CO remove this reconstruction and force the Pt(100) surface to return to the square symmetry and the accompanying lower density of the underlying crystal [74, 75].

The images in Figs. 6.11, 6.12 and 6.13 indicate that there was again a clear difference between structures observed at high and low pressures. Whereas the Pt(100) surface was still unreconstructed at high CO pressures, the *hex*-reconstruction appeared to return when the surface was exposed to a high partial pressure of NO. Even though the resolution in the STM images was insufficient to distinguish the platinum atoms and, thus, to identify these two surface structures directly, the images present several clues that allow us to draw conclusions about these structures. First of all, the height variations in the height images, for both CO-rich and NO-rich gas mixtures, all correspond to the interlayer distance d_{100} of Pt(100), thus showing that the surface has precisely the same structure everywhere; oxides with more than one thickness, such as we have observed on various platinum and palladium surfaces, would have given rise to a variety of height differences that are not integer multiples of d_{100}. Second, the 20 % difference in areal density between the two structures leads to the expulsion of 20 % of a monolayer of platinum atoms when the *hex*-reconstruction of the surface is lifted. This leads to the sudden formation of platinum islands that cover 20 % of the substrate, when the gas mixture is made sufficiently rich in CO. Similarly, when the ratio is made NO-rich, the surface reconstructs and forms vacancy islands that accommodate the 20 % deficit of

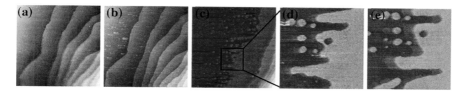

Fig. 6.11 STM images taken with the Mark I ReactorSTM during NO reduction by CO on Pt(100) at a total pressure of 1.25 bar and a temperature of 395 K. **a** Flat terraces under NO-rich conditions (90 %). The height variations visible in the image all correspond to the step height on Pt(100). **b** During the transition from NO-rich to CO-rich conditions. **c–e** Several minutes later in a CO-rich atmosphere (75 % and higher). Islands are formed on the terraces and the steps become wavy. After its introduction, this roughness decays with time. (**a–c** 400 x 400 nm, **d–e** 120 x 120 nm). (After [73])

Fig. 6.12 STM images taken with the Mark I ReactorSTM during NO reduction by CO on Pt(100) some time after switching from a CO-rich gas mixture to an NO-rich atmosphere (85 %). The total pressure was 1.25 bar and the temperature was 382 K. The initial surface, in the CO-rich mixture, had been smooth. The switch to the NO-rich mixture had resulted in the formation of vacancy islands in the platinum surface. This roughness decayed with time, with the vacancy islands disappearing and the steps becoming less wavy. (**a** 310 x 300 nm, **b–d** 120 x 120 nm, **e** 350 x 370 nm). (After 73)

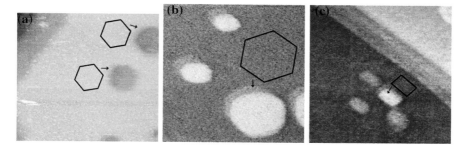

Fig. 6.13 STM images taken with the Mark I ReactorSTM during NO reduction by CO on Pt(100) briefly after changes in gas composition. Two different symmetries are found for the adatom or vacancy islands. **a, b** Under NO-rich conditions a weak hexagonal symmetry can be distinguished. **c** CO-rich conditions correlate with a rectangular symmetry. (After 73)

platinum in the outermost atom layer. In both cases the adatom or vacancy islands evolve via a combination of Smoluchowski ripening (coalescence dynamics) and Ostwald ripening (atom-by-atom exchange between islands), which makes them disappear after some time, thus restoring a flat surface with a low step density.

The third clue is that when the mixture is switched from NO-rich to CO-rich and the *hex*-reconstruction is lifted, we temporarily observe a significant influence of the STM tip on the surface structure: the adatom islands that are formed line up precisely in the direction of motion of the tip. This suggests that immediately after the phase transition, there is a temporarily high density of highly mobile platinum adatoms on the surface that accumulate into islands under the influence of the modest Van der Waals attraction exerted by the tip. After a few minutes the islands disappear and the tip influence effect is no longer observed. Finally, as can be seen in Fig. 6.13, the shapes of the vacancy islands formed upon switching to NO-rich conditions, rounded hexagons, were different from those of the adatom islands that were formed upon switching in the opposite direction, squares and rectangles.

Of course, when the surface structure changes, it is obvious that also the binding energies for atoms and molecules on the surface should change and so should the energy barriers for adsorption, desorption, dissociation and for the reactions between the adsorbed species. Measurements with the quadrupole mass spectrometer of the production of N_2, obtained simultaneously with the STM images, are consistent with Langmuir–Hinshelwood kinetics, both for CO-rich and for NO-rich gas mixtures. The fit to the experimental data becomes slightly better when we allow each rate constant in the fit to have two different values, one for the CO-dominated, unreconstructed surface and the other for the NO-dominated, *hex*-reconstructed surface. The most important difference that we find between the two sets of best-fit rate constants is that the ratio of the rate constants for NO adsorption and desorption seems to be higher on the *hex*-reconstructed surface than on the unreconstructed surface [73]. This makes sense, in view of the fact that the correspondingly increased adsorption strength actually stabilizes the *hex*-reconstruction under these conditions. The increase in rate constant is larger than the 20 % increase in the number of potential adsorption sites, which indicates that the effect is, at least partly, due to the difference in electronic structure between the reconstructed and unreconstructed surfaces.

6.5 Conclusion and Outlook

The ReactorSTM is providing us with a direct view of working model catalysts under realistic or nearly realistic conditions, involving the combination of high pressures and elevated temperatures. The Mark II version of the instrument, which has recently become operational, images the model catalyst surfaces under reaction conditions with full atomic resolution at high imaging rates, even above 1 frame/s [76, 77]. This instrument is now in a sufficiently mature state that a commercial version has been developed in the meantime by Leiden Probe Microscopy B.V [58].

The examples given here of the oxidation of CO and the reduction of NO on catalytically active metal surfaces demonstrate the pressure gap between atmospheric pressures and the traditional regime of low-pressure experiments, where simple adsorption structures and the familiar Langmuir-Hinshelwood reaction mechanism prevail. The effect of the high pressure is to thermodynamically (or kinetically) stabilize new surface structures and compositions that incorporate one or more of the reactants. This is introducing a much more active role of the catalyst. Instead of serving merely as a two-dimensional arena on which the reactants adsorb, encounter each other and react, with each of these steps affected by the interaction with the substrate, the catalyst surface is involved much more intimately in the process at higher pressures; one might rather describe its role as that of an intermediate product, providing a truly alternative reaction pathway. This scenario is not necessarily limited to CO oxidation. Similar effects may well be at play in a wider variety of catalytic processes, in which the partial pressures of one or more of the reactants (or products) may be high enough to introduce a phase transition to a new structure of the catalyst surface, involving e.g. oxides, carbides, nitrides, sulfides or other compounds.

It is clear that the complexity introduced by the high pressures in these experiments on well-defined, single-crystal, elemental catalysts can be taken as a guarantee that the materials gap, between these model systems and the full richness of a supported catalyst, containing a support material, small particles and promoters, will have important surprises for us in store. Adapting our techniques to make them suitable for more realistic catalyst structures and compositions therefore forms a very relevant challenge. A first step in this direction is the development of a non-contact Atomic Force Microscopy (nc-AFM) version of our instrument, the ReactorAFM. A prototype of such an instrument, with the nc-AFM running by means of a miniature tuning fork, has recently been constructed. First results obtained with the ReactorAFM show that it is possible to perform in-situ observations during catalytic reactions on non-conducting specimens, e.g. on model systems of supported catalysts in the form of metal nanoparticles on an insulating, oxide support.

A limiting factor in both the ReactorSTM and the ReactorAFM and in other high-pressure scanning probe instruments is the quality of the tips that are used to probe the surface. The tip is exposed to the same gas environment as the catalyst surface. Typical tip materials, such as tungsten and platinum-iridium, are sensitive to the presence of reactive gas mixtures. Adsorption certainly also takes place on the tip and at sufficiently high oxygen pressures, the tip may even oxidize. Even though such tip changes usually do not lead to a resistance that disables the STM, they can have a severe impact on the resolution obtained with the microscope. More work will be necessary to identify materials and manufacturing methods, required to produce reliable and sharp tips that are not affected by the typical conditions used in the experiments.

In closing, we should emphasize that 'seeing' a catalyst at work with the ReactorSTM or the ReactorAFM will at best provide part of the picture. For a full understanding of catalytic processes under practical conditions it will be essential to combine a wide spectrum of experimental techniques, involving real-space and reciprocal space geometrical information, spectroscopic information and providing access to the ultra-fast timescales of the actual bond-breaking and -making processes, all under the appropriate high-pressure, high-temperature conditions. Several high-pressure techniques have been mentioned already in Sect. 3. In this respect, the flow-reactor setup that we have developed recently for surface X-ray diffraction under reaction conditions at atmospheric pressures deserves special attention [78]. Finally, for a proper, in-depth interpretation and eventually prediction of the experimental findings on catalysts under reaction conditions, it is essential that the experimental progress go hand in hand with further development of appropriate theoretical and computational tools [9–12].

Acknowledgments This chapter would not have been possible without the work by many people. In particular, I would like to mention Kees Herbschleb, Dunja Stoltz, Marta Cañas-Ventura, Violeta Navarro-Paredes, Sander Roobol, Bas Hendriksen, Stefania Bobaru and Peter Rasmussen. A special thanks goes out to the technical support staff of Leiden University, in particular to Peter van der Tuijn who has been responsible for the design of the latest version of the ReactorSTM. I am indebted to Gertjan van Baarle and Alexei Ofitserov of Leiden Probe Microscopy B.V., who have combined the development of a commercial version of the ReactorSTM [58] with a significant role in the development of the prototype instruments in the author's laboratory. This project is financially supported by a Netherlands SmartMix grant and the NIMIC partner organizations (http://NIMIC.Physics.LeidenUniv.nl) through NIMIC, a public–private program.

References

1. J. Wintterlin, Adv. Catal. **45**, 131 (2000)
2. P. Stoltze, J.K. Nørskov, Phys. Rev. Lett. **55**, 2502 (1985)
3. J.A. Rodriguez, D.W. Goodman, Surf. Sci. Rep. **14**, 1 (1991)
4. H. Over, Y.D. Kim, A.P. Seitsonen, S. Wendt, E. Lundgren, M. Schmid, P. Varga, A. Morgante, G. Ertl, Science **287**, 1474 (2001)
5. B.L.M. Hendriksen, J.W.M. Frenken, Phys. Rev. Lett. **89**, 046101 (2002)
6. K. Reuter, M. Scheffler, Phys. Rev. B **65**, 035406 (2001)
7. K. Reuter, M. Scheffler, Phys. Rev. Lett. **90**, 046103 (2003)
8. K. Reuter, M. Scheffler, Phys. Rev. B **68**, 045407 (2003)
9. K. Reuter, D. Frenkel, M. Scheffler, Phys. Rev. Lett. **93**, 116105 (2004)
10. J. Rogal, K. Reuter, M. Scheffler, Phys. Rev. B **75**, 205433 (2007)
11. J. Rogal, K. Reuter, M. Scheffler, Phys. Rev. Lett. **98**, 046101 (2007)
12. J. Rogal, K. Reuter, M. Scheffler, Phys. Rev. B **77**, 155410 (2008)
13. H. Bluhm, M. Hävecker, A. Knop-Gericke, M. Kiskinova, R. Schlögl, M. Salmeron, MRS Bull. **32**, 1022 (2007)
14. F.G. Requejo, E.L.D. Hebenstreit, D.F. Ogletree, M. Salmeron, J. Catal. **226**, 83 (2004)
15. W.P.A. Jansen, A.W. Denier van der Gon, G.M. Wijers, Y.G.M. Rikers, H.H. Brongersma, P.W. van der Hoogen, J.A.M. de Laat, T.M. Maas, E.C.A. Dekkers, P. Brinkgreve, Rev. Sci. Instrum. **73**, 354 (2002)
16. P.L. Gai, E.D. Boyes, S. Helveg, P.L. Hansen, S. Giorgio, C. Henry, MRS Bull. **32**, 1044 (2007)
17. P.L. Hansen, J.B. Wagner, S. Helveg, J.R. Rostrup-Nielsen, B.S. Clausen, H. Topsøe, Science **295**, 2053 (2002)

18. A. Stierle, A.M. Molenbroek. MRS Bull. **32**, 1001 (2007), and other review articles in this special issue, dedicated to the theme of *Novel In Situ Probes for Nanocatalysis*
19. J.F. Creemer, S. Helveg, G.H. Hoveling, S. Ullmann, A.M. Molenbroek, P.M. Sarro, H.W. Zandbergen, Ultramicroscopy **108**, 993 (2008)
20. E. de Smit, I. Swart, J.F. Creemer, G.H. Hoveling, M.K. Gilles, T. Tyliszczak, P.J. Kooyman, H.W. Zandbergen, C. Morin, B.M. Weckhuysen, F.M.F. de Groot, Nature **456**, 222 (2008)
21. G. Rupprechter, MRS Bull. **32**, 1031 (2007)
22. G.A. Beitel, A. Laskov, H. Oosterbeek, E.W. Kuipers, J. Phys. Chem. **100**, 12494 (1996)
23. E. Ozensoy, C. Hess, D. Loffreda, P. Sautet, D.W. Goodman, J. Phys. Chem. B **109**, 5415 (2005)
24. X. Su, P.S. Cremer, Y.R. Shen, G.A. Somorjai, Phys. Rev. Lett. **77**, 3858 (1996)
25. G. Rupprechter, T. Dellwig, H. Unterhalt, H.-J. Freund, J. Phys. Chem. B **105**, 3797 (2001)
26. H.H. Rotermund, Surf. Sci. **386**, 10 (1997)
27. J. Evans, A. Puig-Molina, M. Tromp, MRS Bull. **32**, 1038 (2007)
28. A. Knop-Gericke, M. Hävecker, T. Schedel-Niedrig, R. Schlögl, Catal. Lett. **66**, 215 (2000)
29. S. Ferrer, M.D. Ackermann, E. Lundgren, MRS Bull. **32**, 1010 (2007)
30. H. Isern, K. Peters, P. Steadman, O. Robach, J. Alvarez, E. Lundgren, S. Ferrer, Surf. Sci. **482–485**, 101 (2001)
31. P.B. Rasmussen, B.L.M. Hendriksen, H. Zeijlemaker, H.G. Ficke, J.W.M. Frenken, Rev. Sci. Instrum. **69**, 3879 (1998)
32. B.L.M. Hendriksen, *Model Catalysts in Action—High-Pressure Scanning Tunneling Microscopy* (Ph.D. Thesis, Leiden University, 2003), ISBN 90-9016597-5
33. C.T. Herbschleb, *ReactorSTM—Imaging Catalysts under Realistic Conditions* (Ph.D. Thesis, Leiden University, 2011), ISBN 978-90-8593-098-3
34. B.J. McIntyre, M. Salmeron, G.A. Somorjai, Rev. Sci. Instrum. **64**, 687 (1993)
35. J.A. Jensen, K.B. Rider, Y. Chen, M. Salmeron, J. Vac. Sci. Technol. B **17**, 1080 (1999)
36. E. Lægsgaard, L. Österlund, P. Thostrup, P.B. Rasmussen, I. Stensgaard, F. Besenbacher, Rev. Sci. Instrum. **72**, 3537 (2001)
37. A. Kolmakov, D.W. Goodman, Rev. Sci. Instrum. **74**, 2444 (2003)
38. M. Rößler, P. Geng, J. Wintterlin, Rev. Sci. Instrum. **76**, 023705 (2005)
39. F. Tao, D. Tang, M. Salmeron, G.A. Somorjai, Rev. Sci. Instrum. **79**, 084101 (2008)
40. J.A. Jensen, K.B. Rider, M. Salmeron, G.A. Somorjai, Phys. Rev. Lett. **80**, 1228 (1998)
41. E. Kruse Vestergaard, P. Thostrup, T. An, E. Lægsgaard, I. Stensgaard, B. Hammer, F. Besenbacher, *Phys. Rev. Lett.* **88**, 259601 (2002)
42. P. Cernota, K. Rider, H.A. Yoon, M. Salmeron, G.A. Somorjai, Surf. Sci. **445**, 249 (2000)
43. K.B. Rider, K.S. Hwang, M. Salmeron, G.A. Somorjai, Phys. Rev. Lett. **86**, 4330 (2001)
44. K.B. Rider, K.S. Hwang, M. Salmeron, G.A. Somorjai, J. Am. Chem. Soc. **124**, 5588 (2002)
45. P. Thostrup, E. Kruse Vestergaard, T. An, E. Lægasgaard, F. Besenbacher, J. Chem. Phys. **118**, 3724 (2003)
46. B.J. McIntyre, M. Salmeron, G.A. Somorjai, J. Vac. Sci. Technol. A **11**, 1964 (1993)
47. L. Österlund, P.B. Rasmussen, P. Thostrup, E. Lægsgaard, I. Stensgaard, F. Besenbacher, Phys. Rev. Lett. **86**, 460 (2001)
48. A. Kolmakov, D.W. Goodman, Surf. Sci. **490**, L597 (2001)
49. M. Rössler, S. Günther, J. Wintterlin, J. Phys. Chem. C **111**, 2242 (2006)
50. R.T. Vang, J.-G. Wang, J. Knudsen, J. Schnadt, E. Lægsgaard, I. Stensgaard, F. Besenbacher, J. Phys. Chem. B **109**, 14262 (2005)
51. E.K. Vestergaard, R.T. Vang, J. Knudsen, T.M. Pedersen, T. An, E. Lægsgaard, I. Stensgaard, B. Hammer, F. Besenbacher, Phys. Rev. Lett. **95**, 126101 (2005)
52. M. Montano, M. Salmeron, G.A. Somorjai, Surf. Sci. **600**, 1809 (2006)
53. A. Menzel, K.-C. Chang, V. Komanicky, Y.V. Tolmachev, A.V. Tkachuk, Y.S. Chu, H. You, Phys. Rev. B **95**, 035426 (2007)
54. Y. Jugnet, F.J. Cadete Santos Aires, C. Deranlot, L. Piccolo, J.C. Bertolini, Surf. Sci. **521**, L639 (2002)

55. M.S. Hoogeman, D. Glastra van Loon, R.W.M. Loos, H.G. Ficke, E. de Haas, J.J. van der Linden, H. Zeijlemaker, L. Kuipers, M.F. Chang, M.A.J. Klik, J.W.M. Frenken, Rev. Sci. Instrum. **69**, 2072 (1998)
56. M.J. Rost, D.A. Quist, J.W.M. Frenken, Phys. Rev. Lett. **91**, 026101 (2003)
57. G. Dong, E.B. Fourré, F.C. Tabak, J.W.M. Frenken, Phys. Rev. Lett. **104**, 096102 (2010)
58. Leiden Probe Microscopy B.V., (Leiden, Netherlands), www.LeidenProbeMicroscopy.com
59. E. Vlieg, I.K. Robinson, K. Kern, Surf. Sci. **233**, 248 (1990)
60. T. Gritsch, D. Coulman, R.J. Behm, G. Ertl, Phys. Rev. Lett. **63**, 1086 (1989)
61. B.L.M. Hendriksen, S.C. Bobaru, J.W.M. Frenken, Top. Catal. **36**, 43 (2005)
62. STM movies can be viewed on, ip.physics.leidenuniv.nl
63. C. Quiros, O. Robach, H. Isern, P. Ordejón, S. Ferrer, Surf. Sci. **522**, 161 (2003)
64. M.D. Ackermann, T.M. Pedersen, B.L.M. Hendriksen, O. Robach, S.C. Bobaru, I. Popa, C. Quiros, H. Kim, B. Hammer, S. Ferrer, J.W.M. Frenken, Phys. Rev. Lett. **95**, 255505 (2005)
65. G.A. Somorjai, *Introduction to Surface Chemistry and Catalysis* (Wiley-Interscience, New York, 1994)
66. P. Mars, D.W. van Krevelen, Chem. Eng. Sci.http://www.sciencedirect.com/science/article/pii/S0009250954800054 **3**, 41 (1954)
67. S.C. Bobaru, *High-pressure STM studies of oxidation catalysis* (PhD Thesis, Leiden University, 2006), ISBN 90-9021165-9
68. B.L.M. Hendriksen, S.C. Bobaru, J.W.M. Frenken, Surf. Sci. **552**, 229 (2004)
69. B.L.M. Hendriksen, S.C. Bobaru, J.W.M. Frenken, Catal. Today **105**, 234 (2005)
70. R. Westerström, J.G. Wang, M.D. Ackermann, J. Gustafson, A. Resta, A. Mikkelsen, J.N. Andersen, E. Lundgren, O. Balmes, X. Torrelles, J.W.M. Frenken, B. Hammer, J. Phys. Cond. Mat. **20**, 184018 (2008)
71. R. Westerström, J. Gustafson, A. Resta, A. Mikkelsen, J.N. Andersen, E. Lundgren, N. Seriani, F. Mittendorfer, M. Schmid, J. Klikovits, P. Varga, M.D. Ackermann, J.W.M. Frenken, N. Kasper, A. Stierle, Phys. Rev. B **76**, 155410 (2007)
72. B.L.M. Hendriksen, M.D. Ackermann, R. van Rijn, D. Stoltz, I. Popa, O. Balmes, A. Resta, D. Wermeille, R. Felici, S. Ferrer, J.W.M. Frenken, Nat. Chem. **2**, 730 (2010)
73. C.T. Herbschleb, S.C. Bobaru, J.W.M. Frenken, Catal. Today **154**, 61 (2010)
74. A. Hopkinson, D.A. King, Chem. Phys. **177**, 433 (1993)
75. P. van Beurden, Phys. Rev. Lett. **90**, 066106 (2003)
76. M.J. Rost, L. Crama, P. Schakel, E. van Tol, G.B.E.M. van Velzen-Williams, C.F. Overgauw, H. ter Horst, H. Dekker, B. Okhuijsen, M. Seynen, A. Vijftigschild, P. Han, A.J. Katan, K. Schoots, R. Schumm, W.A. van Loo, T.H. Oosterkamp, J.W.M. Frenken, Rev. Sci. Instrum. **76**, 053710 (2005)
77. J.W.M. Frenken, T.H. Oosterkamp, B.L.M. Hendriksen, M.J. Rost, Mater. Today **8**, 20 (2005)
78. R. van Rijn, M.D. Ackermann, O. Balmes, T. Dufrane, A. Geluk, H. Gonzalez, H. Isern, E. de Kuyper, L. Petit, V.A. Sole, D. Wermeille, R. Felici, J.W.M. Frenken, Rev. Sci. Instrum. **81**, 014101 (2010)

Chapter 7
Detectors for Electron and X-ray Scattering and Imaging Experiments

Alexander Ziegler and Heinz Graafsma

Abstract Suitable detectors for these expensive and highly complex experimental instruments described in the previous chapters are a key factor to consider, primarily because if one cannot visualize or record the experimental results with an appropriate detector, any experiment will fail. The general challenge for all position-, energy-, and time-resolving detector systems is the fulfillment of stringent requirements for direct X-ray and electron detection experiments. These include a priori a high detection sensitivity and efficiency, but most important is coping with extremely high flux (10^{12} highly energetic X-ray photons or 10^8 300 kV electrons per second), exhibiting appropriate radiation hardness to maintain proper detection sensitivity and operability, low electronic noise for finest energy resolution in single-photon counting mode, and high frame rates for high time resolution. Parameters such as the Modulation Transfer Function (MTF), the Detector Quantum Efficiency (DQE), the dynamic range, pixel size, sensitivity, linearity, uniformity, background noise, read out speed, and reliability (or life time) among other characteristics will need to be considered to decide which detector design is best for what application. There are a variety of designs in the development and/or prototype stage. Costs are high, because most are produced using expensive wafer fabrication processes. A point of consideration is flexibility, adaptability, and how

A. Ziegler (✉)
Microscopy and Microanalysis Unit, The University of the Witwatersrand,
1 Jan Smuts Ave., Johannesburg 2000, South Africa
e-mail: alexander.ziegler@wits.ac.za

H. Graafsma
Photon-Science Detector Group,
Deutches Elektronen-Synchrotron, DESY, Notkestr. 85, D-22607 Hamburg, Germany
e-mail: heinz.graafsma@desy.de

H. Graafsma
STC Research Center, University of Mid-Sweden, 85107 Sundsvall, Sweden

A. Ziegler et al. (eds.), *In-situ Materials Characterization*,
Springer Series in Materials Science 193, DOI: 10.1007/978-3-642-45152-2_7,
© Springer-Verlag Berlin Heidelberg 2014

swift detector parameters can be changed. The trend at high-end, multi-national, multi-user scientific research facilities (Synchrotrons, FELs) however, is to operate dedicated, non-transferable detectors for specialized applications, whereas the medium to small scale research facilities may well decide for a more versatile, multi-purpose detector. The following sections will address detectors for electrons and detectors for X-ray photons separately. Development efforts for these detector types overlap, in part due to the high costs involved, and in part due to the compatibility of some developmental stages and components for both detector types.

7.1 Detectors for Electrons

7.1.1 Introduction

Detectors for electron imaging have continuously been evolving over the past three decades. Before the 1980s photographic film was the only choice for suitable electron image recording. The transition from the very traditional photographic film over imaging plates, indirect detection techniques, to the novel direct detection techniques has provided electron microscopists with a number of options for image recording. In brief, the traditional photographic film comes with the associated light sensitive negatives, their development in the dark room and printing on photographic paper. The imaging plates use less light sensitive negatives, but still require digitization and printing. The indirect detection technique using Charge Coupled Devices (CCD) require electron-to-light conversion methods—i.e., scintillators and CCDs that operate in the visible range of the spectrum. Finally, the novel digital detectors use direct conversion of the incoming particle, electron or photon, to an electrical signal and perform digitization in the detector front-end.

The motivation for the transition from the classical photographic film over imaging plates to electronic, CCD-based, indirect electron detectors is the sheer advantage of having a digital image, which is directly and immediately viewable and adjustable on a computer screen at the time of recording. This is certainly more user friendly than having to deal with negatives and their developing conditions (and the capabilities of the microscopist), and the post-microscope session scanning of negatives or image plates. The biggest disadvantage of photographic film or the imaging plate, is that mishaps and system failures can often only be seen after the negatives have been developed or the imaging plate read out. The possibilities that digital imaging offer—not only in terms of direct image acquisition, but also for any ensuing on-line and/or live image processing—have opened doors to a number of research fields, in particular but not exclusively, insitu microscopy, electron tomography, and remote microscopy (where the operator may be located in an adjacent office or even in a different city or country). Digital detectors have therefore become the de-facto standard in electron optical imaging.

However, none of the electron imaging devices is perfect. From the plain image recording perspective each device has advantages and disadvantages that need to be taken into account depending on the application.

7.1.2 Electron Image Recording Devices

7.1.2.1 Photographic Film

Electron microscopy conform photographic film consists of an emulsion of fine silver halide particles (3–6 μm in size) suspended (1:1) in an approximately 20 μm thick layer of gelatine. The film thickness corresponds approximately to the mean free path of 120 kV electrons—purposefully designed so that these electrons scatter statistically only once while traversing the film. Multiple scattering inside the film layer would only degrade the quality and the resolution of the recorded image, because lateral scattering and propagation of the primary electron away from the incident direction would only generate metallic silver particles where no incident electron may have hit the film. The silver halide grains convert to metallic silver grains when exposed to the electron beam. The energy required for this conversion is approximately 7 eV, clearly indicating that one single high-energy electron in an electron microscope can easily generate 10^4 silver atoms—representing a large portion of one single silver halide grain in the film emulsion. This process stores a latent image in the film made of silver halide grains. In comparison, film exposure to visible light of about 3 eV requires a collective process involving many photons to convert the silver halide atoms to metallic silver atoms. During chemical development of the photographic film in the dark room, the silver grains containing the latent electron image information are fully converted to metallic silver, whereas the non-irradiated and therefore non-converted silver halide grains are removed.

For photographic film, its non-linearity, the limited possibilities to reduce the inherent noise characteristics, and the limited dose (or intensity) range, plus the strong dependence on the film developing conditions, make film not quite suitable for applications like low-dose imaging, or where electron intensity quantification is important. One has for decades recorded electron diffraction patterns on film, and it has shown very suitable for the recording of high-intensity peaks (except for the central diffraction peak). However, with the non-linear characteristics of film, and its best operating conditions in an intermediate yet limited dose (intensity) range, film tends to reduce the true intensity of the very high-intensity peaks, appropriately represent the medium-intensity peaks, and obscure the low-intensity peaks below recognition. Precise quantification of the electron intensities however, is important for convergent beam electron diffraction patterns. With film, this is done by measuring the optical density in a film scanner. The optical density here is defined as the fraction of light that is transmitted by the film. However, considering the non-linearity of film the quantification allows reconstructing the true intensities only when knowing the non-linear relation between optical density and dose. Systematic errors affect the results significantly. Nevertheless, in some cases experimental and theoretical intensities can be matched with an accuracy of close to 99 %.

7.1.2.2 Imaging Plate

An imaging plate consist of an approximately 110 μm thick photo-stimulable layer of phosphor grains (BaFX {X = Cl, Br}, doped with Eu^{2+}) that is encapsulated by a flat sheet of plastic support below and a protective layer above. The grains are nominally regarded as pixels 25 × 25 μm in size. This flexible sheet can be used just like film in the standard film cassette of the microscope. During recording, the phosphor layer stores the electron energy in the potential well of the defect states of the phosphor grains. In essence, by ionizing Eu^{2+} to Eu^{3+}, some of the electrons are excited to the conduction band and are trapped in F^+-centers of the crystal creating a metastable state. This process allows the phosphor grains to store a latent image. However, at room temperature these defect states do have a limited life time of up to a few days, during which one can digitally read out the imaging plates many times. However, each read out reduces the amount of energy stored, and as such the signal decreases as the number of read outs increases. The read out mechanisms relies on the narrow spectral range of a He–Ne laser that scans the imaging plate causing luminescent light emission from the imaging plate that is captured, logarithmically amplified by a photomultiplier, and digitized into 14 bits. Exposing the imaging plate to white light erases the latent image by irradiating the phosphor with sufficient photon energy, causing the removal of the defect states in the phosphor grains and a back-conversion of the Eu^{3+} ions to Eu^{2+}, such that the imaging plate can be re-used. All imaging characteristics are the combined response of the imaging plate and the read out device. One cannot isolate the characteristics of the imaging plate from the reader.

Imaging plates exhibit a highly linear counts versus electron dose behavior, offer possibilities to reduce the inherent noise characteristics by adjusting the read out device to a high sensitivity setting, and their broad dose (or intensity) range in which they operate, make imaging plates very suitable for low-dose and high-resolution imaging applications. They are not very good at recording and storing high electron doses (intensities) as the inherent noise characteristics increase dramatically with dose. This signal-to-noise limitation is due to granular variations of the phosphor. The 25 × 25 μm pixel of an image plate consist of a varying number of Eu^{2+} doped phosphor grains (due to grain size variations). Consequently, the number of Eu atoms varies from pixel to pixel, and concomitantly the gain factor, which in return causes large variations in gain across the image plate, contributing to the intensity dependent noise, also known as granular noise. In principle—and this is technically feasible—the read out device would need to be precisely aligned with the image plate inserted, such that it normalizes these gain variations. For this to work out properly, the positioning and alignment error of the image plate would need to be less than 1 μm for every read out. Another remedy to reduce the gain variations among pixels is to reduce the grain size of the phosphor, such that a more homogeneous distribution of grains (and Eu atoms) can be achieved. Binning of pixels is another method that allows reducing gain variations, however at the expense of spatial resolution.

7.1.2.3 CCD-Based Indirect Electron Detector

The working principle of the CCD-based, indirect detection strategy involves three processes, namely (1) capturing and converting electrons to photons via a phosphor scintillator, (2) guiding these photons to a light sensitive CCD array via fiber optical coupling, and (3) converting the photons back into an electronic signal via electron-hole pair generation in the CCD pixel, which can be read out consecutively to form an image.

This indirect electron detection approach is characterized by a number of optical interfaces within the entire detector assembly, resulting in multiple scattering events and a subsequent loss in resolution. The scintillator layer at the very top of the detector assembly consists of a material that converts electrons to photons. A variety of powder phosphors, in particular Gadolinium Oxysulphide (GOS) phosphor, as well as an Yttrium Aluminum Garnet (YAG) single crystal doped with Cerium serve this purpose. Both materials have significantly different photon yields. The generation of photons in GOS phosphor is about twice as high as in YAG, resulting in a highly sensitive detector if equipped with GOS phosphor. Yet, the variations in GOS powder morphology, especially grain size, cause a larger variance in response characteristics. In contrast, the YAG crystal exhibits a more uniform photon emission and is more resistant to electron beam irradiation—hence a favorite material for CCD based electron detector manufacturers. Additionally, scintillator thickness affects electron-to-photon conversion—the gain. A thicker layer emits more photons than a thinner scintillator, allowing the tailoring of detector sensitivity according to the application. Generally, low-dose applications call for the enhanced gain found in thicker scintillators, because the sample consists of beam sensitive, low-contrast matter (biological samples), and a decent signal-to-noise ratio is critical. Instead, a thin scintillator is used for high-dose applications, examining more beam resistant samples that exhibit better contrast, and aiming at high spatial resolution in the image.

While thicker scintillators yield higher gain, the result is not independent of the variable accelerating voltage of the electrons in electron microscopes (in most cases anywhere between 30 and 300 kV). Lower accelerating voltages and thus slower incident electrons have a shorter mean free path—that is the distance between consecutive scattering events—than higher accelerating voltages (and faster electrons). Consequently, a slow electron will scatter more often in a given scintillator thickness, yielding a higher gain than a fast electron. This is shown in Fig. 7.1 where the dependence of gain on accelerating voltage is plotted for a CCD-based electron detector in comparison to an imaging plate.

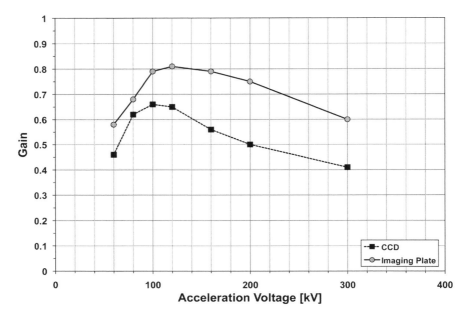

Fig. 7.1 The dependence of detector gain on accelerating voltage—comparing a CCD-based electron detector to an imaging plate

The CCD-based detector is equipped with a fiber-coupled YAG scintillator, whereas the imaging plate pixels consist of phosphor powder. Observe how the gain peaks at 100–120 kV accelerating voltage and drops off at higher accelerating voltages. The scintillator thickness of this particular detector shown, as well as the imaging plate was designed to yield highest gain for 100–120 kV mircoscopes/ applications. (Note: 100–120 kV or even 60 kV applications are not to be confused with low-dose applications. Low-dose microscopy can also be performed at 300 kV. Low-dose only stands for low-intensity, not low-accelerating voltages.)

As mentioned before, it can also be observed that the YAG crystal is less gain efficient than the phosphor used in the imaging plate.

Electron scattering occurs in all scintillators, resulting in photon generation across a number of phosphor grains, or alternatively, across a number of optical fibers that guide the photons to the CCD pixel array. Fiber-coupling connects each pixel via an optical fiber with the scintillator. Thus, one single electron that ideally should activate only one corresponding pixel on the CCD may in reality—due to scattering—activate some neighboring pixels too. While a thicker scintillator may generate a larger number of photons, it also exhibits a larger spatial spreading of these photons over a larger number of optical fibers. Thus spatial resolution in the recorded image is compromised. Additionally to the spreading of photons, reflection of photons back into the YAG scintillator (from both, the entry and exit plane) contributes to lateral signal spreading and image blur. Photons that enter the optical fibers at high incident angles can—provided the internal reflection conditions are such that the acceptance cone (critical angle) is not too small—cause a leaking of photons out of fibers adding to the image blur and loss of spatial resolution.

An alternative to fiber-coupling the scintillator with the CCD pixel array is lens-coupling. Lens-coupling allows for a somewhat higher flexibility in detector assembly and design, but it is less efficient than fiber-coupling. Fiber-coupling is done by permanently attaching a set of parallel fibers (bound together forming a block of fibers) to the scintillator and an identical block of fibers to the CCD. When assembling the detector, the fiber blocks from the scintillator and the CCD need to be aligned properly and connected optically. This is best done via a thin film of oil spread between the blocks of optical fibers. In any case, the measured overall response of the detector is a combined response of scintillator, fiber optics, CCD array, and the interfaces therein.

Compromised spatial resolution due to electron scattering and the ensuing photon propagation in the optical fibers is the major limiting factor for the CCD based electron detectors. Although the individual pixel in the CCD array is small (the majority of CCD have 14 or 24 μm pixels), this is rarely the actual image resolution as usually more than one pixel is activated by one single incident electron.

Very high electron intensities, encountered for example in the central spot of a diffraction pattern, over saturates the CCD—even if blocked with the beam stop of the electron microscope, the brightness "spills over", causing oversaturation of an entire array of CCD pixel rows and columns. There are some CCD based detectors that can handle higher intensities, with the consequence that they do not perform well in the medium to low electron intensity regime.

Beside the image quality compromising effects in the scintillator and fiber optics, the CCD based electron detectors are also characterized by significant levels of conversion, readout and dark-current noise in the pixel array. The dark current and the readout noise (from here on denoted as background noise) are proportional to the temperature of the pixel array and the thermally excited electrons inside each pixel are proportional to the exposure time. Experimentally they are hard to distinguish. This background noise can be reduced by cooling the detector down to −25 °C—usually with Peltier cooling elements. Both, the readout and the dark current noise are independent of illumination, dominating in low-dose images, whereas the photon-to-electron conversion noise in the pixel increases with increasing illumination. The latter type of noise can strongly affect the signal-to-noise ratio at high-dose applications. However, it can be minimized by performing a gain normalization of the entire detector area prior to image recording. To understand this, consider each pixel as an independent detector. Hence, the CCD array consists of many small, separate detectors, each one with its own set of background and conversion noise. The conversion noise in each of them results in a distinctive gain response that increases proportional to the illumination intensity, and each individual gain response varies constantly during continuous exposure. By recording a number of full sized (entire detector array) images under uniform illumination, these gain variations can be averaged and normalized to adjust the pixel-to-pixel variations across the entire pixel array. Similarly, one can perform a dark current assessment by recording a number of dark current images (no illumination), which for obvious reasons include the readout noise. Both, the resulting gain normalization and the dark current images are subtracted from each image that is recorded during the ensuing microscope session. In contrast, this type of digital image processing for noise reduction purposes cannot be performed with photographic film or the imaging plate.

The CCD-based indirect electron detectors exhibit a highly linear counts versus electron dose behavior, and their broad dose (or intensity) range (also called dynamic range) in which they operate, make these digital detectors very suitable for medium- to high-dose applications. However, considering the disadvantages of the indirect electron detection strategy, especially the inferior spatial resolution due to relatively large pixels, and the lack of sensitivity in the low-dose domain due to the high levels of inherent background noise, are motivation enough to continue with electron detector development.

Thus, the reason and motivation to further transition from the indirect to the next level—direct electron detection techniques—is the prospect to finally operate an electron detector with all the digital and user friendly advantages of the existing CCD based detector systems, yet with better characteristics. In particular, the ability to investigate highly beam sensitive samples, especially in the soft-matter and the life sciences is sufficient motivation to advocate further detector development. The direct electron detection technique has the potential to provide a much more sensitive imaging device, because its noise and spatial resolution figures can be up to 10 times better. Single electron detection is the goal here. These enhanced capabilities of direct electron detectors have inimitable consequences, because it allows taking on the major problem in beam sensitive investigations using electrons: radiation damage. Radiation damage to the sample is also found in experiments using other high-brightness radiation sources, i.e., synchrotron and X-ray Free Electron Lasers (XFEL). Fact is that the ultimate resolution of structural studies using electrons and/or X-rays is limited by the radiation sensitivity of the sample.

There are in principle two approaches to overcome the degrading effects of radiation damage. One is to perform single-shot experiments, where a large number of electrons ($>10^8$) or X-ray photons ($>10^{12}$) is directed in one single, but very short pulse (10–1000 fs) onto the specimen. The idea here is to record the image (or spectral information) faster than the damage mechanism can unfold and destroy the sample. Generally, the damage mechanisms can be associated with the ionization of atoms due to the incoming X-rays. Once ionized and exhibiting a charge, the subsequent effect of so many same-charge ions, which are still densely packed in their original volume causes a so-called Coulomb explosion. Such ultrafast experiments can be performed and they are currently of high scientific interest (see Chaps. 2 and 4), concomitant with the development of XFELs. The second approach pertains so far only to the use of electrons and entails performing single-electron experiments allowing for energy and heat dissipation within the sample and thus avoiding thermal damage mechanisms to destroy the sample. Both approaches have in common that they require highly sensitive direct detectors. Additionally, the very large dynamic range of these novel detectors becomes very important here.

In order to be able to objectively compare electron imaging devices one needs to understand the critical parameters that are generally used to characterize detectors. The concepts of spatial resolution (pixel size) and the Modulation Transfer Function (MTF), noise and the Detector Quantum Efficiency (DQE), dynamic range, linearity, and uniformity are the parameters of interest. While we will discuss all of them in the following section, the MTF, the DQE, and the dynamic range, are the primary parameters used to compare electron detector types here. There is an increasing demand and an enormous list of publications (too many to list them all here) that all aim at uniform and standardized detector characterization methods. However, due to the large variability of detector designs and applications, this task is not as easy and straightforward as it may seem [1]. The medical imaging community has their standards, for instance the IEC-62220-1 for the determination of the DQE. Applying this standard to similar yet different detectors in other disciplines proves practically impossible in many cases.

7.1.3 Critical Parameters for Detector Characterization

7.1.3.1 Modulation Transfer Function

The MTF measures the ability of a detector to distinguish a black-white transition. Ideally, a black-and-white variable-width line pattern (Fig. 7.2) would need to be placed right in front of the detector and an image recorded, and even more ideally in both the horizontal and the vertical direction. The ability of any image recording device to perfectly record and display such black-white features will decrease (and thus the MTF decline) as the features—here the line spacing—become smaller and smaller. Eventually, the black-white-black will turn into shades of grey and the ability to discern accordingly small features in a high-resolution image will vanish.

However, such an ideal fine line pattern cannot be fabricated with line spacings small enough (<1 nm) to cover the entire detector's ability to transfer spatial frequencies. As one alternative to such a fine line pattern only one black-white transition—a sharp knife-edge—is used to record images for the MTF analysis. A second alternative is the noise approach. In the latter a series of noise images are recorded and analyzed to extract the required information for the MTF.

In essence, the MTF measures how contrast is transmitted within the detector. Contrast is defined as

$$Contrast = \frac{I_{max} - I_{min}}{I_{max} + I_{min}} \tag{7.1}$$

where I_{max} and I_{min} are the intensity maxima and minima respectively. Figure 7.3 shows a simulation of the effect the MTF has on a set of parallel lines that vary in separation (increasing distance from 1 to 11). The top row (a) depicts the ideal image without blurring of the signal (perfect contrast, perfect MTF = ideal

Fig. 7.2 A *black-and-white* variable-width *line pattern* used to determine the modulation transfer function

detector). The rows below (b–i) show an increasing signal blur (real detector) in conjunction with decreasing contrast and a decreasing MTF (j).

The reason for a detector to not be able to fully reproduce such line patterns with razor-sharp edges in the image is the spreading or blurring of the signal inside of the detector. The photographic film, the imaging plate and the scintillator/fiber optic/CCD array they all exhibit the inherent scattering events of electrons and photons. Accordingly, one single electron (and the ensuing photons), which enters the detector at one single point, excites not only one single pixel or grain but the signal spreads to neighboring pixels and grains too. This behavior can be expressed by the Point Spread Function (PSF), which describes the response of the detector to a single incident electron or photon—the input signal gets convoluted with the detector response, producing the output signal. If one were to illuminate one single pixel or grain with one single electron one could directly determine the PSF. However this method is extremely challenging. Two more manageable approaches are to follow the sharp knife-edge method or the noise method. Both methods assume that the PSF is independent of the position of the pixel in the array.

The knife-edge method requires placing a very sharp metallic edge across of the detector area. The material should not permit electron transmission. The resulting input signal into the detector can therefore be described by a step-function. However, the detector output signal shows a blurred edge in the image. Note though, that the knife-edge method has one-dimensional character—just like the step-function—not allowing determination of the PSF in the other lateral direction (along the knife-edge). Hence, a Line Spread Function (LSF) is used instead of the PSF. The LSF can be determined as the first derivative of the step function, or from the experimental edge profile, the edge spread function (ESF).

$$LSF = \frac{d}{dx}ESF(x) \approx \frac{\Delta ESF}{\Delta x} \tag{7.2}$$

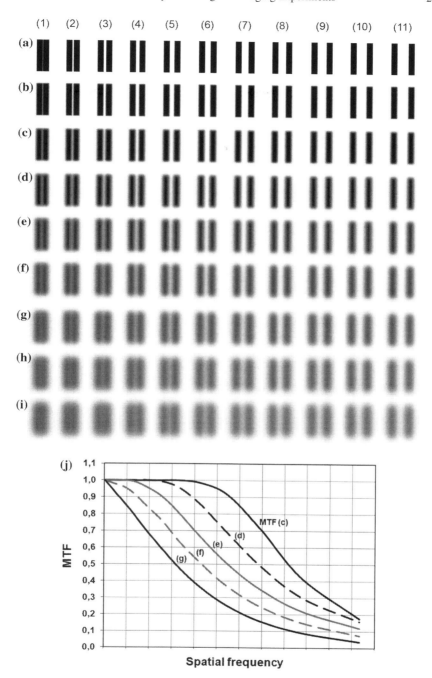

Fig. 7.3 Image simulation of the effect the MTF has on features in the image. Row **a** shows the unaffected image (perfect detector), rows **b–i** demonstrate increasing blur, **j** depicts the corresponding MTF graphs used for the image simulation

The ESF is determined by recording the knife-edge image, averaging over each line (or column) of pixels that runs parallel to the knife-edge and normalizing. The ESF cannot be differentiated analytically, hence the finite difference method applies.

Unfortunately, direct differentiation of the experimental edge profile has shown to be too noisy for a decent assessment of the LSF. Alternatively, a fitting procedure can be employed. The analytical approach to the LSF is by superimposing Gauss functions. The ideal input signal (perfect step function) is then convoluted with this analytical LSF and the result is compared to the experimental output signal. Variation of the free parameters of the Gauss functions allows adjusting the analytical LSF to match the experimentally measured LSF. The LSF corresponds to the PSF only under the condition that the PSF has rotational symmetry. However, in a real detector the PSF can be anisotropic. In photographic film this effect is known as Eberhard effect [2]. However, once the PSF (or the LSF) has been determined experimentally, the MTF is acquired by taking the Fourier transform of the PSF (LSF).

The issue of the noisy edge function and therefore a rather ineffective LSF can be explained by the metallic knife-edge causing stray X-rays and diffracted electrons right at the edge that can reach the detector where it should not be illuminated. This effect contributes to the artificial blurring of the edge that is independent of the detector response. Thus, the knife-edge method tends to overestimate the spreading of the input signal. On the other hand, inserting a sample into the electron beam might just as well introduce this type of edge blur—obviously, depending on the material composition. While the knife-edge is supposed to be inserted right above the detector to minimize such scattering effects, the sample is usually inserted at a much higher position along the electron beam. Thus, minute scattering along, for example, the ion-milled hole in the sample or any thicker section, and which may be regarded as considerably less scattering than the knife-edge, can cause similar effects in the image, which is recorded at a much larger distance. Thus, it remains debatable, whether or not it is better to de-convolute an experimental image (with sample) with an MTF that was acquired without image distorting scattering, than using an MTF that was captured with a possibly more realistic experimental scenario. Reproducibility and standardization of such distinct test methods are a challenge though.

An alternative to the knife-edge method to determine the MTF of a detector is the noise method. It requires a homogeneous illumination of the entire detector area. The number of electrons per unit time and unit area that reach the detector are Poisson distributed. An ideal detector would therefore have pixel/grains that are statistically independent (each a small individual detector, with no PSF effects) and the frequency spectrum of the noise could be expressed by a constant. This number would only depend on the variance of the noise in the image. In a real detector, however, the recorded homogeneous noise image requires a convolution with the PSF and the inherent detector noise (background noise, conversion noise) needs to be added.

$$Noise\ image = Poisson\ noise \otimes PSF + detector\ noise \qquad (7.3)$$

Discrete Fourier Transformation (*DFT*) yields the frequency spectrum.

$$DFT(Noise\ image) = DFT(Poisson\ noise) \times DFT(PSF)$$
$$+ DFT(detector\ noise) \qquad (7.4)$$

where $DFT(PSF) = $ MTF, and $DFT(Poisson\ noise) = $ constant + noise.

As can be seen the PSF can be calculated by taking the Fourier back transform of the MTF. The constant is unknown and as such the result has to be scaled and normalized to unity under the assumption that for small spatial frequencies the MTF tends to unity. Averaging over the entire illuminated area allows minimizing the noise in the frequency spectrum. This can be achieved by first Fourier transforming and then averaging the noise line-by-line or column-by-column. This method also allows examining the rotational symmetry of the MTF and the PSF. Any additional inherent noise of the detector introduces only systematic errors to this method. Therefore, to minimize these errors one needs to record these noise images in a dose (intensity) range where the inherent detector noise is negligible compared to the noise of the input signal.

7.1.3.2 Detector Quantum Efficiency

The DQE is a most commonly used measure characterizing the overall quality of an imaging detector. Imaging quality is particularly affected by the combined effects of signal and noise performance expressed as a function of object detail—spatial frequency. In other words, it determines how well a detector can distinguish important low-contrast information from noise. This is of particular importance for low-dose applications, where the number of incident electrons is limited, e.g., due to sample radiation exposure limits, and resulting beam damage. Figure 7.4 demonstrates how image contrast and noise interplay.

Any signal inherently comes with its noise spectrum—call it shot noise, which should be Poisson distributed. An ideal detector registering such signal under perfect imaging conditions, will transfer that signal (and its noise) to the read-out end without adding any further noise (background noise) to the signal—hence, it has a DQE = 1. However, in practice the detector background and/or conversion noise will cloud the true signal, making it difficult, for example, to clearly discern very small features of interest in the image. A detector cannot increase the true signal, but it always adds noise—consequently, it has a DQE < 1. Background and conversion noise are inescapable in any digital imaging chain. Thus, image quality strongly depends on the often used Signal-to-Noise Ratio (SNR).

It is intuitive that without input signal there is only uniform background noise that can be displayed. Increasing the signal to, for example, a few electrons (or photons) may by far not suffice to allow distinguishing the signal from the background noise either. Increasing the signal more and more may result in a threshold signal strength that will eventually permit distinguishing the signal from the background noise. There are a number of threshold values of what the minimum SNR needs to be for the signal to qualify as a 'true' or 'good' signal. SNR numbers of 3:1 and 5:1

Fig. 7.4 Demonstration of
how contrast and noise affect
image quality

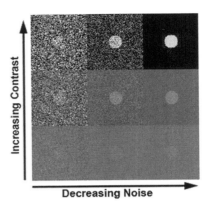

are very common ratios. Hence, the minimum number of incident electrons required
to meet a specified SNR is a measure of a certain level of image quality. Knowing
(ideally) also the noise level that was involved in this specific SNR allows determining the number of incident electrons that would have been sufficient for this image
quality level if the noise had been non-existent—in essence the number of equivalent
incident electrons, K. Relating K to the actual number of incident electrons, q, that
were used to record an image with specified SNR defines the DQE

$$DQE = \frac{K}{q} \tag{7.5}$$

In the ideal case where noise is non-existent, $K = q$ and the ideal detector is characterized by DQE = 1. A real detector, as mentioned above, will have DQE < 1. In
other words, the DQE describes how good a real detector represents the true image
information relative to the ideal detector.

Experimentally, it is a challenge to determine both K and q with high accuracy.
Even with very sensitive electron current measurements in the pico-Amp range,
1 pA still corresponds to over 6 million electrons, and at this level the measurement error is as large a the value measured.

Alternatively, the DQE can be expressed as [3]

$$DQE = \frac{SNR_{out}^2}{SNR_{in}^2} = \frac{\left(\frac{I_{out}^2}{\sigma_{out}^2}\right)}{\left(\frac{q^2}{\sigma_{in}^2}\right)} \tag{7.6}$$

relating the signal-to-noise ratios of the output to the input signal. However, in this form
care is advised as this equation is only valid if the input signal corresponds to a uniform
Poisson distribution. In the second expression I_{out} is the detector read out count rate, q
the number of incident electrons (as before), and σ_{out} and σ_{in} are the respective vari-
ances. Assuming uniform Poisson distribution, $q = \sigma_{in}^2$ simplifying (7.6) to

$$DQE = \frac{I_{out}^2}{q\sigma_{out}^2} \tag{7.7}$$

Further simplification can be arranged by considering the conversion rate, c. Each detector system associates a count rate (usually counts per second) to a specified number of incident electrons (or electron intensity). That count rate is what the detector displays, for example as a pixel value (or grey scale value). The conversion rate, c, is therefore defined as the count rate per incident electron.

$$c = \frac{I_{out}}{q} \tag{7.8}$$

A CCD-based electron detector allows determining and adjusting this conversion rate at the A/D converter. Depending on the detector manufacturer conversion rates vary between 5 and 50. Combining (7.7) and (7.8) gives

$$DQE = c\frac{I_{out}}{\sigma_{out}^2} \tag{7.9}$$

Remember that adjacent pixels cannot be considered as completely isolated from each other and that the PSF tends to spread the incident image information over a few neighboring pixel. This results in a noise-leveling operation. The input noise may therefore be higher than the output noise, causing an unrealistic DQE > 1. This error can be compensated by introducing a mixing factor [4]

$$\sigma_{out,\,comp}^2 = \frac{\sigma_{out}^2}{m} \tag{7.10}$$

which is defined as [5]

$$m = \frac{1}{N^2}\sum_{i=1}^{N}\sum_{j=1}^{N} MTF_{i,j}^2 \tag{7.11}$$

where N is the number of pixels in one dimension. This mixing factor is independent of the incident dose. For an ideal detector $m = 1$, whereas for a real detector $m < 1$. Therefore, (7.9) needs to be amended

$$DQE = c\frac{mI_{out}}{\sigma_{out}^2} \tag{7.12}$$

The above equations allow determining the DQE experimentally by selecting a specified region of interest, for example, 512×512 pixel taken from noise images recorded under varying electron intensities.

With

$$I_{out} = \frac{1}{512^2}\sum_{i=1}^{512}\sum_{j=1}^{512} I_{i,j}$$

and

$$\sigma_{out}^2 = \frac{1}{512^2}\sum_{i=1}^{512}\sum_{j=1}^{512}\left(I_{i,j} - I_{out}\right)^2 \tag{7.13}$$

and (7.11), one can experimentally determine the DQE using (7.12).

Note, that the mixing factor, m, introduces the MTF into the DQE equation. This allows a direct correlation of the DQE with the spatial frequencies in the image. In other words, if a sample is viewed under the same contrast and detector noise conditions, large features of the sample can be deciphered easier than small features in the recorded image. This makes the DQE a rather important parameter, more than for example, the spatial resolution limit of a microscope. Even if a microscope can in theory resolve features on the sub-nanometer scale, one cannot take advantage of those resolution capabilities if the detecting system has low DQE, which prevents recognizing very small objects in the image. Consequently, high DQE (i.e., high SNR) or low system noise is therefore key to capturing the greatest portion of useful image information. Often, the only way to compensate for poor SNR is to increase radiation dose, which is for many applications an unacceptable trade-off.

There is a connection between DQE and CCD pixel size. Generally, the larger the pixel size the better the SNR. This is because the input signal is much better converted into output signal in a large pixel, while the pixel noise remains constant. A minimized PSF also contributes to this improvement as the incident electrons—although scattering inside the larger pixel as much as before—do not cause a spreading of the incident signal over a number of neighboring pixels. As a logical consequence, CCD-based electron detectors should preferably display large pixels. However, from the spatial resolution point of view small pixels are better suited for high-resolution imaging. Hence, a careful evaluation of application requirements is mandatory when designing an imaging detector. There is a point of diminishing return: while noise remains constant in a given detector system, the amount of useful signal captured per pixel dwindles with decreasing pixel size. As a result, the finer the pixel matrix, the lower the SNR at each pixel. This, combined with the lower inherent contrast of small objects will limit detectability and visualization. Determining the optimum balance between pixel size and noise is therefore crucial to the development of CCD-based electron detectors.

The ideal case of single-electron recording can be best achieved with large pixels (>100 μm). Capturing electron after electron, suitably spaced in time as they arrive at the pixel and are read out individually before the next electron enters the pixel is the best for single-electron detection, because the SNR is optimized under such conditions. The draw back of miserable spatial resolution with such large pixels can be overcome by employing algorithms designed to back-track the trajectory of the electron to its point of entry into the pixel. These algorithms rely on scattering simulations and center of gravity determination of the 'scattering cloud' and the energy deposited by the electron. This method works only if the scattering cloud covers two or more pixels. Equipped with such algorithms one can virtually reduce the pixel size to <10 μm and thus have a more accurate positioning of the electron in detector-space. However, in real electron imaging more than just one electron at a time is registered by each pixel. Already two incident electrons will make the back-tracking and positioning algorithms infinitely more difficult to handle. Any large number of electrons will render this back-tracking approach highly impractical, if not impossible when considering that normal electron imaging deals with electron numbers, per pixel, per exposure (<500 ms) in the 10^6 range.

7.1.3.3 Dynamic Range

The dynamic range of a detector is defined as the range between the largest and smallest possible values of a variable quantity—in this case the number of electrons that a detector can register. This range should be as large as possible, from a single electron to several millions of electrons in one and the same image—a condition encountered for example in diffraction patterns. For the CCDs behind the scintillators that convert electrons to light, the dynamic range of such CCDs is many times less than that of the human eye (in the visible spectral range). Assessing the dynamic range for film with electrons is difficult, due to the ambiguity on how to determine the minimum and maximum detectable signals. For CCD-based electron detectors the minimum is certainly determined by the readout noise of the detector.

7.1.3.4 Linearity

The linearity of a detector is generally determined by recording images of uniform illumination and plotting detector readout (counts) as a function of electron intensity (electron dose). However, the difficulty lies in accurately determining the electron intensity on both the high- and low-dose end. Linearity is important when intensity variations are to be determined as is the case for structure factor determination.

7.1.3.5 Uniformity

The uniformity of a detector measures the gain variations among different pixels. Examining the raw data (pixel values) of an image that was recorded under uniform illumination will reveal that there are large variations in the readout signal among pixels. As every pixel has its own readout pathway each one will therefore exhibit unique conversion and gain characteristics. It is not unusual with CCD-based electron detectors that neighboring pixels differ by a factor of 10 in gain, thus, the almost identical incident signal (considering the PSF effects) results in a pixel that may be ten times brighter (or dimmer) than its neighbor. On a larger scale, this behavior can even manifest in entire detector areas that respond stronger than other areas of the detector to the identical uniform illumination. The ideally uniform image may now look as if it were displaying stains.

Similar gain variations but not as distinct, applies to film and the imaging plate. Variations in gain depend here on the phosphor grain morphology and orientation relative to the incident electron that may cause stronger or weaker scattering events inside affecting energy deposition and structural changes (silver halide to silver conversion). Regardless of random grain orientation and a homogeneous distribution, neighboring phosphor grains may differ significantly in their response to electron irradiation.

In CCD-based electron detectors this non-uniformity can be corrected for by averaging over many images recorded with the same uniform illumination. This primarily levels out gain variations over time within each pixel but not among neighboring pixels or any other pixel in the detector area. The average image gain, $\bar{g}_{img}$, is calculated according to

$$\bar{g}_{img} = \frac{\bar{I}_{out} - \bar{b}}{\bar{N}_e}$$ (7.14)

where $\bar{b}$ is the average background noise and $\bar{N}_e$ is the averaged electron dose per pixel. The gain variations among all pixels are reduced by taking the average pixel gain and dividing it by the average image gain.

$$\gamma(n) = \frac{\bar{I}_{out}(n) - \bar{b}(n)}{\bar{N}_e \bar{g}_{img}} = \frac{\bar{I}_{out}(n) - \bar{b}(n)}{\bar{I}_{out} - \bar{b}}$$ (7.15)

7.1.3.6 Resolution and Pixel Size

For a discrete 2-dimensional detector the spatial resolution obtained in the image is solely determined by the pixel size. However, in practice a detector's spatial resolution is determined by the PSF as discussed before, and not simply by the detector pixel size. The problem arises when the PSF extends as much as 70–90 μm FWHM—as generally obtained in commercial CCD-based electron detectors. In light of that, all efforts to improve the spatial resolution in the image by reducing the pixels size as much as technically possible (standard pixel size is currently 14 μm, with some detectors having 5 μm pixels) will not be as successful as expected. As long as the PSF extends beyond the pixel boundaries and cannot be confined to it, the PSF will be determining the spatial resolution. Numerous optical elements and interfaces as encountered in CCD-based electron detectors contribute to signal spreading and a broad PSF. However, knowing the particular PSF of a detector allows deconvoluting each image, and in return improving the spatial resolution in the image. This possibility explains why it is not customary to settle for pixel sizes as large as the PSF for a specific detector.

As mentioned already at the end of Sect. 7.1.3.2, pixel size has to be carefully chosen, depending primarily on application. Single-electron and ultra-low-dose applications will benefit from large pixels and their improved SNR characteristics. Instead, regular and high-dose applications will most likely not benefit from large pixels because of the lack of spatial resolution in the image.

7.1.4 Comparison of Electron Detectors

The following section will present a few results of those critical detector characterization parameters for the 'conventional' electron detectors. Figure 7.5 shows

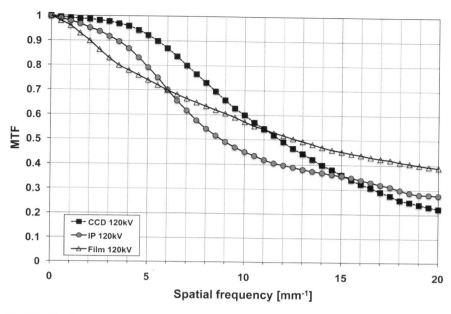

Fig. 7.5 The MTF is plotted as a function of spatial frequency, f, for photographic film (Kodak 4463), an image plate (Fuji IP25), and a CCD-based electron detector (Gatan 679 with YAG scintillator). The noise method was used for the determination of the MTF

the MTF plotted as a function of spatial frequency, f, for photographic film (Kodak 4463), an image plate (Fuji IP25), and a CCD-based electron detector (Gatan 679 with YAG scintillator). The film has the typical 3 μm silver halide grains, whereas the pixels sizes of the image plate and the CCD-based electron detector are 25 and 24 μm respectively. Consequently with 25 μm pixel, the MTF shown here is limited to the frequency range 0–20 mm^{-1}—up to the Nyquist frequency of the image plate, which is defined as $f_N = 1/(2 \cdot \text{pixel size})$. Furthermore, the regions of interest in all three detectors were chosen to be 1024 × 1024 pixel (grains) in size.

The noise method was used for the determination of the MTF. However, due to the sensitivity of this method to inherent background noise, and in the attempt to reduce the negative effects caused by background noise, the MTF was determined in the medium- to high-dose range for the CCD-based electron detector and the photographic film, while the image plate was analyzed using low-dose electrons. Remember that reproducibility and standardization of such distinct MTF test methods are a challenge. The described methods of analysis used for determining the MTF of a detector, especially their computational aspect may vary only little in their approach, but may produce considerably different looking graphs. Although a uniform and consistent analysis method is everybody's aspiration and intention, differences in, for example, the Gauss fitting parameters, the line and/or column averages, and especially the order in which these separate steps are applied to the data set can make a significant difference in the outcome. Thus, in publications

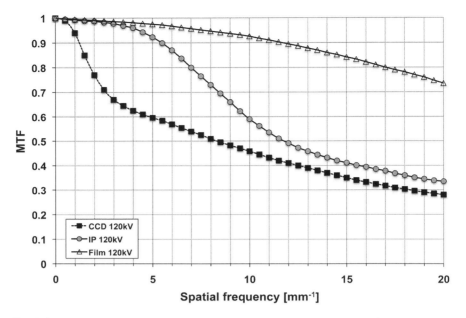

Fig. 7.6 Alternative MTF graph, plotted as a function of spatial frequency for film, image plate, and CCD-based electron detector

one can find MTF graphs as presented in Fig. 7.5, but also MTF graphs (even of the same detector) that look like the graph shown in Fig. 7.6.

The MTF graph displayed in Fig. 7.5 shows how the difference between the three detection systems can be minimal over a wide range of spatial frequencies. Only at low spatial frequencies, the CCD-based electron detector is performing better than the other two detection systems, whereas at high spatial frequencies the photographic film is transmitting information better, i.e., 0.38. The MTF of all three detection systems in this high spatial frequency domain agree well with other publications [6–8] where CCD-based electron detectors achieve MTF values of 0.22–0.28, and imaging plates are slightly better with 0.27–0.33.

Alternatively, Fig. 7.6 shows how the CCD-based electron detector now relays the least amount of information, whereas the image plate and the photographic film are better in this regard [9]. Across the entire range of spatial frequencies the photographic film is the best recording media. Film still transmits 0.73, i.e., 73 %, of the original signal at 20 mm^{-1}, whereas the image plate and the CCD-based electron detector transmit only 0.34 and 0.28 respectively. The value for the CCD-based electron detector and the image plate at 20 mm^{-1} agrees well with what is presented in Fig. 7.5, however, the shape of the curve and its behavior at low spatial frequencies is considerably different.

Transmission of information at high spatial frequencies is generally desirable for high-resolution electron microscopy work, so the MTF behavior of the detection system is of significant interest there.

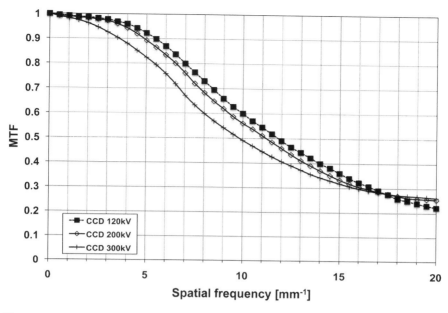

Fig. 7.7   The effect of electron acceleration voltages on the MTF

The effect of electron acceleration voltages (120, 200 and 300 kV) on the MTF is shown in Fig. 7.7. The MTF tends to be better at lower acceleration voltages, especially at the low-spatial frequency end. The reason for this behavior is the diminished scattering of the incident electron in the scintillator and therefore the signal is spread less over the fiber-optics and the ensuing pixels.

Figure 7.8 shows the DQE of these three detector types. The DQE is shown as a function of electron dose in terms of electrons per unit area (μm^2), because the pixel and grain sizes differ. Furthermore, it is a double logarithmic graph. The DQE of the CCD-based electron detector exhibits a fairly high and level DQE in the medium- to high-dose range with values around 0.45–0.55. This relatively high DQE for CCD-based electron detectors agrees well with other published investigations [5, 6]. The DQE however decreases to <0.2 in the low-dose range, i.e., at less than 0.006 e/μm^2, which is caused by the inescapable background noise in the CCD that becomes predominant at low electron doses.

The imaging plate and film show larger variations. The imaging plate has a high DQE in the low-dose range, 0.4, and decreases in the medium- to high-dose range to levels below 0.07. This is due to the conversion noise that increases considerably with electron dose. Film instead, shows a decent DQE only in the medium-dose range, 0.38, and drops off steeply at high and low doses. The general shape and position of the DQE for film can be manipulated (within narrow limits) by varying the film development parameters. The results agree with the observation that a CCD-based detector with high DQE can improve one's ability to distinguish

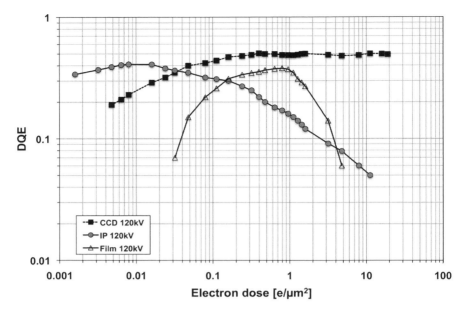

Fig. 7.8 DQE of three detector types, shown as a function of electron dose in terms of electrons per unit area (μm^2)

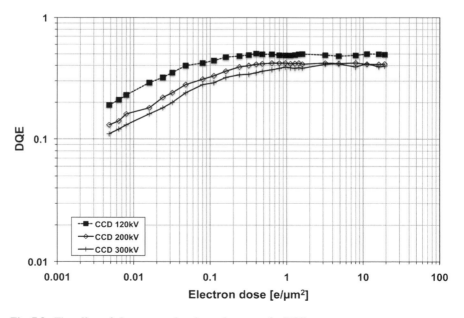

Fig. 7.9 The effect of electron acceleration voltages on the DQE

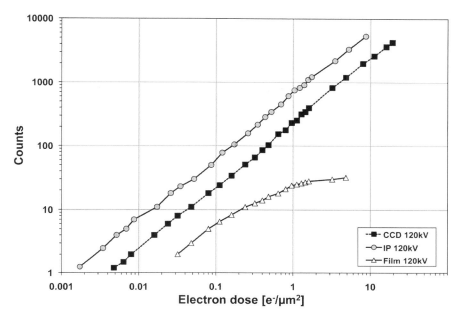

Fig. 7.10 The linearity of the three detector types: CCD based electron detectors, film, and image plate

small, low-contrast features from background noise—even though this high-DQE detector may have substantially inferior spatial resolution capabilities, i.e., significantly larger pixel size, than film.

Similar to the MTF, one can also notice an effect of acceleration voltage on the DQE performance of the CCD-based electron detector. Figure 7.9 shows how the DQE changes by varying the acceleration voltage (120, 200 and 300 kV).

Considering the dynamic range, and linearity: the CCD-based electron detector and the imaging plate make full use of their analog-to-digital converter that provides for 16 and even 32 bit. Film instead is limited to approximately 11 bit (depending on the film scanner). The linearity of the three detector types is shown in Fig. 7.10. Note the large scale of the electron dose (more than 10 $e/\mu m^2$, which translates to approx. 6000 e/pixel at 24 and 25 μm pixel size) that the CCD-based electron detector and the imaging plate can handle, while still exhibiting a linear response. Film instead is highly non-linear and not as a capable in handling large electron doses.

7.1.5 Direct Electron Detectors

As mentioned at the end of Sect. 7.1.2.3, the reason and motivation to further transition from the indirect to the direct electron detection techniques, is the prospect to finally operate an electron detector with all the digital and user-friendly advantages of the existing CCD based detector systems, yet with improved characteristics. The main driving force for these detector developments are the high-energy physics and the medical imaging communities. Therefore, the applications differ significantly as well as the type and energy range of the radiation (high and low energy electrons and X-rays).

Direct detection generally entails that the detector is directly exposed to the radiation source (electrons, X-rays, visible photons, ions, and other ionizing particles). Electron-to-photon conversion is not required anymore, which eliminates the problem of an extended PSF caused by optical interfaces within a CCD-based detector system (see Sect. 7.1.2.3). The direct exposure to the incident electron beam however poses a different set of challenges that need to be overcome, radiation hardness being the most 'damaging' problem.

7.1.5.1 The Active Pixel Sensor

The development of direct detections technique dates back as far as 1968 and 1969 [10–12], even before the CCD was invented in 1970 at Bell Labs. However, technical difficulties in the manufacturing process of direct detectors made the CCD more viable and applicable at the time. The direct detection techniques back then and today are all based on the Complementary Metal-Oxide-Semiconductor (CMOS) technology for designing and manufacturing integrated circuits (IC). In the 1990s the CMOS technology was in a better position to reliably manufacture ICs at smaller and smaller scales, allowing for the design of a direct detector that used intra-pixel charge-transfer possibilities [13]. These CMOS-based direct detectors are called Active Pixel Sensor (APS). The basic APS design and its functionality will be described in the following section. There are a large number of design possibilities and layout variations one could dive into—too many to cover them all in the limited the space provided here. Furthermore, many of these detector variations and designs are still in the prototype stages with extensive and costly development, improvement, and testing stages still ahead.

A CMOS type APS, designed to efficiently detect electrons, X-rays, and other ionizing particles, especially the low energy type of radiation, differs from any other standard CMOS imaging sensor. The critical parameter is the thickness of a p-epitaxial layer that determines detector sensitivity. A very thin layer allows only very few secondary electrons (electron-hole pairs) to be created via scattering events of the primary, incident electron, causing a very shallow junction depth, and consequently the amount of collected charges inside the sensing layer is limited. As a result the signal-to-noise ratio (SNR) of such a detector is poor. Increasing the epitaxial layer thickness improves the SNR significantly.

Figure 7.11a demonstrates in a schematic cross section of the detector the working principle of the direct detection mechanisms. It shows how an incident, primary electron creates secondary electrons through scattering events along its path throughout the entire layered structure of the detector. The $p++$ bulk layer is

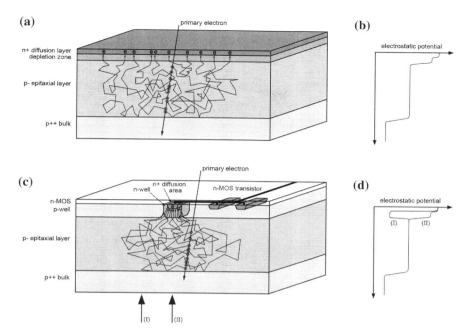

Fig. 7.11 A schematic cross section of (**a**) a CMOS detector and (**b**) an APS detector, depicting the working principle of the direct detection mechanism. **c** and **d** show their respective electrostatic potentials

a highly doped standard silicon wafer that acts as support structure. A p-epitaxial layer is grown on top of the substrate and is the key feature because it is an integrated epitaxial silicon layer (purer silicon of higher resistance) that acts as the sensing component of a detector. On top of that is an $n+$ diffusion layer, which is added as a standard practice in normal CMOS processing. A depletion zone is formed at the interface between the $n+$ diffusion layer and the p-epitaxial layer, which is equivalent to an $n+/p-$ diode. This layered structure creates internal electric fields and potential barriers and gradients that allow manipulation and guiding of the secondary electrons. For example, the interface between the $p++$ bulk and the p-epitaxial layers create an electric field barrier that reflects secondary electrons back into the epitaxial layer, preventing them from entering the $p++$ substrate. In contrast, in the epitaxial layer there is no electric field, and secondary electrons inside tend to diffuse following a rather random path until they reach the depletion zone. Once inside the depletion zone the secondary electrons experience the electric field of the $n+$ diffusion layer, causing them to drift (rather than to diffuse) into the $n+$ diffusion layer and remain trapped inside its deep potential well. Figure 7.11c depicts the electrostatic potential of such a layered structure as shown in Fig. 7.11a.

The presented concept can be utilized and adjusted to create a highly sensitive sensor, capable of single electron detection. It can also be extended to incorporate the read-out circuitry into the sensor—effectively creating an APS. A schematic cross section through an APS is shown in Fig. 7.11b. The sensor structure shows a similarly layered configuration as before, but also displays the incorporation of additional structures, especially at the top. The $p++$ bulk substrate is the same, and the p-epitaxial layer is thicker than before (not shown in schematic, Fig. 7.11b as it is not to scale), but on top of the epitaxial layer comes a p-well layer with inclusions of lightly doped n-well zones. The former $n+$ diffusion layer is now reduced to small $n+$ diffusion spots within the n-well zones. A depletion zone as before is not specifically shown anymore, but exists in reduced size at the interface between the n-well and the p-epitaxial layer. This configuration is now representing a small diode junction placed in direct contact with the epitaxial layer, and thus capable of collecting all secondary electrons. To enhance electron collection at the small diode and to make the entire epitaxial layer volume and as such the entire pixel surface area sensitive to incoming radiation, the n-well and $n+$ diffusion area is surrounded by p-well structures that reflect the secondary electrons just like bottom $p++$ bulk substrate does, helping to confine and guide the secondary electrons toward the small diode junction. These confining and guiding potential barriers are due to the doping level of the epitaxial layer, which is lower than that of the surrounding p-well structures and the $p++$ substrate, and they are of height [14]

$$V = \frac{kT}{q} \ln \frac{N_{sub}}{N_{epi}} \qquad (7.16)$$

where k is the Boltzmann constant, T is the absolute temperature, q is the electron charge, and N_{epi} and N_{sub} are the doping levels of the epitaxial layer and the substrate, respectively.

An additional benefit of the p-well structures is that they shield the embedded n-channel transistors from the diffusing secondary electrons. Figure 7.11d shows the electrostatic potential at positions (I) and (II) through the cross section of the APS.

The n-channel transistors embedded in the p-well structures and the metallic contacts constitute the on-sensor readout and control electronics. The layout of these electronic components allows for a large variety of configurations. Here, only three basic and very common configurations are shown in Fig. 7.12a–c.

The first one (Fig. 7.12a) represents the very basic version, a Passive Pixel Sensor (PPS), that includes just one single transistor, which multiplexes the collected charge for readout. The layout is very simple and compact, but because it is unbuffered it brings about a slow readout and a poor SNR. The second circuit layout (Fig. 7.12b) represents a standard APS with three transistors (buffering, multiplexing and resetting), resulting in a faster readout and better SNR. Figure 7.12c shows an APS with an additional fourth transistor and a capacitor that can be used for either a shutter function or a sample-and-hold function.

Typical dimensions of these APS layers and structures are: 12–16 μm epitaxial layer thickness, 3 μm square diode, 20 μm square pixel with between 1 and 4

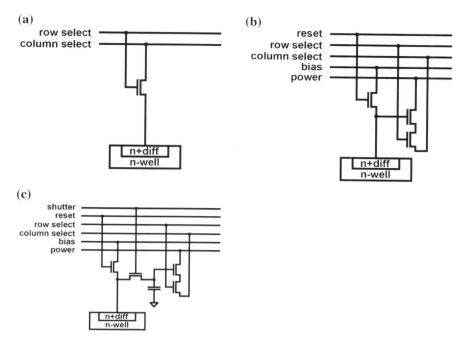

Fig. 7.12 Electric diagrams of three direct detector types, (**a**) a Passive Pixel sensor (PPS), (**b**) a standard Active Pixel Sensor (APS), and (**c**) an APS with additional functionality

diodes. The 4-diode arrangement is expected to reduce charge spreading between neighboring APS pixels. Charge spreading to neighboring pixel is an undesired effect, because it reduces spatial resolution in the recorded image in an identical manner to the CCD-based electron detectors where the signal spreads due to the internal optical interfaces. Unfortunately, the thickness of the epitaxial layer strongly affects this performance. Increasing the epitaxial layer thickness not only increases the number of generated secondary electrons, but it also allows for charge spreading and it increases the collection time, which in turn affects the readout speed. Another structural parameter that needs to be considered when designing an APS is the depth of the n-well zone, as it affects charge collection time. Furthermore, the fill-factor—the ratio of radiation sensitive area to the total area of the pixel—is important. If the electronic circuitry is covering too much surface area of the pixel, it hampers the penetration of primary electrons into the sensitive epitaxial layer. Remember, that for example 300 kV electrons cannot penetrate through samples that are more than 400 nm thick (a typical value for metals), without being severely diverted off the their initial direction of travel—hence becoming useless for TEM imaging purposes. Some of the incoming electrons are absorbed in that 400 nm thick sample, while the rest will exit the sample, but will have lost a significant amount of the energy. The "cloud" of secondary electrons instead can reach much farther into the sample (or detector in this case) and as such the scattering volume is much wider and much deeper than the primary electrons may be able to reach. Hence, covering the surface area of the pixel with circuitry that measures a few micrometers in thickness will make it difficult for the primary electrons to reach the epitaxial layer, whereas the secondary electron cloud may well reach into the epitaxial layer, but generating a weaker signal. X-rays and other ionizing radiation have different penetration depths—again, a clear indication, that the detector requires appropriate design depending on the application.

From the operational aspect, the readout and data handling options are also affecting the performance of an APS. Whether or not the APS pixels in a detector array are reset before each readout (individually or row/column-wise) affects SNR. Increasing the integration or collection time improves SNR.

The bottom line is that detectors with APS offer a large number of parameters that can (and need to) be tuned to the respective application to achieve optimum performance. And it is certainly understandable that most of the current research into APS detectors concentrates on how to improve sensitivity and the SNR, especially for low-energy and low-intensity applications [15–28]. Less research is dedicated to the following problem: radiation hardness [29–32].

Some applications for these APS detectors can be demanding, requiring further development, re-designing and adjustment of current APS detectors. One experimental condition found in numerous investigations is high-intensity and/or high-energy radiation. Such experimental conditions can be detrimental to the APS detector. Damage to the detector can occur in a single, first exposure or over a slightly longer period of time (hours, days, weeks). Therefore, for such applications an APS detector needs to be radiation hard. It has been observed in many APS prototype test experiments that the electronic circuit on top of the APS gets damaged and malfunctions after a short period of time of being exposed to the incoming radiation. The degrading effect is associated with interface charges that hamper detector sensitivity and functionality. One has to remember, that scientists that are primarily interested in the outcome of the experiment may not think about the detector radiation hardness. The radiation source and its intensity may be intentionally selected to destroy the sample in a single, short exposure. In fact, the objective of the experiment may be to observe damage mechanisms. However, what the radiation source can do to the sample it can also do to the detector. In diffraction experiments, for example, there is little one can do to protect the detector from the central, highly intense beam. The only solution here is to manufacture a detector with a hole in the middle, where the damaging central beam can just pass through (and hit a Lead shield behind the detector). In this case, only the electrons that are scattered in the sample are reaching the detector as their path deviates from the central beam. Furthermore, fractions and pieces of the destroyed sample (even in the micrometer size range) that get propelled towards the detector during exposure to the intense incoming radiation can cause physical damage to the detector.

A solution for less intense incoming radiation, which nevertheless may still create damage over time, is to have the detector back-side illuminated. So far the APS detector described here was illuminated from the front, where the circuitry is located. Inverting the detector and removing the p++ substrate (this process is called back-thinning) such that the epitaxial layer is directly exposed to the incoming radiation, opens up another avenue for detector protection and making it radiation hard. The control and readout circuit is now at the back end of the detector and not anymore in the direct "line of radiation fire". For damage to occur to the circuit, the incoming radiation has to traverse the entire epitaxial layer and still possess enough energy to damage the circuit. Depending on application one can increase the epitaxial layer thickness to avoid incoming electrons reaching the backside of the inverted APS (circuit side) altogether. Figure 7.13 shows schematically how an APS detector undergoes back-thinning.

This back-side illumination idea was initially conceptualized to improve sensitivity and SNR, especially for low-energy applications (3–30 kV). As mentioned above, such low-energy electrons would not reach the epitaxial layer if the pixel surface is covered with p-wells and $n+$ diffusion structures a few micrometers in thickness. Even if they were to reach the epitaxial layer, the number of electrons reaching that far and the few secondary electrons created would be too little to generate a sufficiently strong signal that surpasses the electronic background noise within each pixel.

There is an alternative to the APS type of detector: the pn-CCD—or direct CCD—which is a detector based on the CCD working principle, but uses CMOS fabrication processes. While the APS sensor has all functions and all necessary transistors included within one pixel, a direct CCD has those functions remotely operating on a separate circuit board. CMOS image sensors offer superior integration, power dissipation and system size at the expense of system flexibility and image quality, especially in the low-dose range. In contrast, direct CCDs offer superior image quality in the low-dose range, and higher system flexibility, but at the expense of system size. The radiation damage discussed is a problem for APS detectors. In contrast, a direct CCD offers better radiation hardness, because there are no active MOS structures directly exposed to the incoming radiation, and the remaining structures can be designed to be fairly radiation hard.

7.1.5.2 The Direct CCD

The direct CCD comes close to many of the experimental requirements, yet further development is ongoing. To date, this detector performs such that the electronic noise is between 2–5 electrons (rms), frame readout speed is at 1100 frames/s, and single incident photon or electron events can be located to within 5 μm precision. The currently available pixel dimensions of 36–75 μm are optimized for high-speed readout and excellent signal-to-noise characteristics, however, for high-resolution multi-photon/electron imaging (i.e., multiple incidents in one pixel within one readout cycle) the pixel size needs to be reduced to 15–20 um. The concept of this direct CCD is based on a 450 um thick sensitive n-type Silicon layer, which can be fully depleted (sideward) by applying appropriate reverse-biased voltages to the pnp-structure from both the front and back of the direct CCD. Full depletion results in high detection efficiency, because the entire pixel volume is sensitive to radiation.

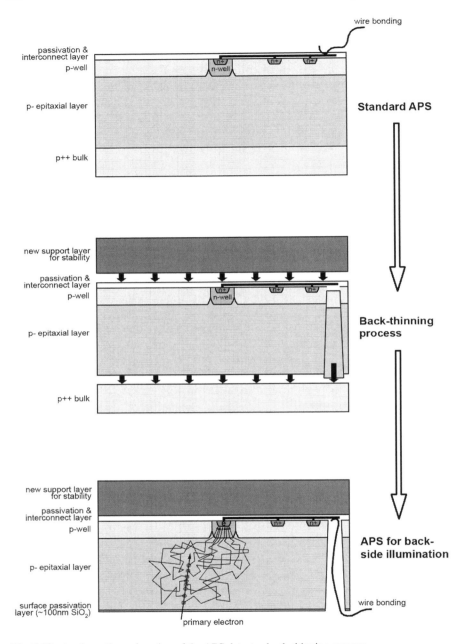

Fig. 7.13 A schematic explanation of the APS detector back-thinning process

By appropriately applying voltages to the *pn*-junctions a potential minimum close to the front surface is created, where electrons generated by the incoming radiation are stored. The pixel structure and transfer registers are created by ion-implantation on the front side, whereas a completely homogeneous and unstructured *p+* implantation on the backside represents the entrance window for the incoming radiation. The rectifying *pn*-junction permits maximizing the electric field at the entrance window, which helps controlling the spatial spread of the charge cloud generated by the ionizing radiation within the pixel. The generated electron-hole pairs (secondary electrons) are then separated owing to their charge and the applied electric field. The electrons drift to their potential minimum, located within a layer of low resistance. During recording, i.e., the photon or electron integration time, the charges are stored within the potential well of each pixel, which is regulated by three transfer registers and two separate voltage levels. The stored charge in the image frame is then quickly transferred to the frame store segment—an adjacent section on the detector, with identical number of smaller and shielded pixels. The image frame can now be exposed again for the next image, while the frame store segment is read out row by row through the readout anodes of each channel of the CCD, amplified, shaped, processed and multiplexed by a dedicated analogue amplifier array, and transferred to an external computer for image processing. The external amplifier chip uses multi-correlated sampling for noise filtering. This chip also controls the detector sampling number, affecting so the detector noise and read out speed characteristics. The sampling number can be adjusted to meet specific experimental conditions. A high sampling number minimizes noise at the expense of read out speed, whereas a low sampling number increases detector speed.

7.1.5.3 Performance of a Direct CCD

A direct CCD detector has been employed on experiments using synchrotron X-rays, electrons, and photons in the visible and near-infrared domain [33–37]. Here, only some characteristics of the direct electron detection will be shown, taken from prototype experiments using a transmission electron microscope (TEM) operated at 100 keV. As mentioned in previous sections of this chapter, one of the parameters of interest to microscopists, and which is primarily relevant for imaging is the MTF.

Figure 7.14 demonstrate the MTF of this direct CCD (*pn*CCD) detector in comparison to a conventional indirect CCD-based electron detector. Clearly, direct detection is superior to conventional indirect detection, because it is still able to record and display high spatial frequency information where the conventional camera fails to do this. This dramatic improvement however, is a direct proof that a reduced number of optical interfaces (no fiber-optics, no scintillator) and thus a reduced number of sites that may cause scattering and signal spreading holds great potential for significant detector improvement.

However, this MTF is the result of ultra-low-dose experiments using electrons. The TEM was adjusted in such manner that the electron dose, incident on the

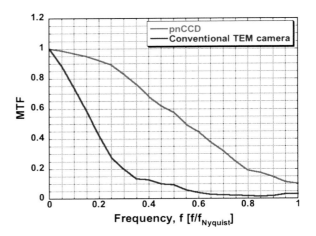

Fig. 7.14 MTF comparing a conventional CCD-based electron detector (indirect detection) to a novel direct electron detector

direct CCD detector, was set to the lowest possible setting. Images were recorded with exposure times of 100 s. A major drawback of such very long exposure times is a compromised spatial resolution in the image due to sample drift (certainly noticeable at high magnifications). Nevertheless, MTF measurements using the knife-edge method could be performed without being negatively affected by sample drift. Furthermore, this ultra-low-dose setting allowed the center of gravity algorithm to be fully exploitable. As mentioned previously, this algorithm relies on electron scattering simulations inside the pixel and the energy deposited by the electron into the pixel. Equipped with such an algorithm the spatial resolution is increased to 5 μm allowing a more accurate positioning of the electron entrance location in detector-space. In practice this single-electron read-out mode enables single electron detection as long as the incident dose corresponds to single-electron illumination conditions.

However, in real electron imaging more than just one electron at a time is registered by each pixel. Detector read-out speed controls the amount of electrons (per pixel per unit time) that can be processed as single-electron events. If the incident dose of electrons exceeds the processing speed, i.e., two or more electrons are registered within one pixel during one read-out cycle, then the back-tracking and positioning algorithms will be infinitely more difficult to handle. Naturally, any large number of electrons that one encounters during 'regular' TEM use will render this back-tracking approach highly impractical.

A small number of prototype detectors are designed to then default to signal integration mode, departing from single-electron counting mode permitting to maintain their operational readiness and expanding their application flexibility.

In conclusion, direct detection systems promise to outperform conventional electron imaging detectors. Some characteristic parameters, which are relevant for imaging, demonstrate this already on a number of prototypes, and there still is room for improvement and fine-tuning of detector fabrication and performance parameters.

7.2 Detectors for X-rays

7.2.1 Introduction

X-ray sources have traditionally been used for the investigation of static structures, rather than dynamic processes using time-resolved or in-situ experiments. There are various reasons for this, and a few of these are listed here without an attempt for completeness. One very important reason is the limited flux at the sample produced by laboratory based X-ray sources, which results in relatively long exposure times. Therefore, only relatively slow processes can be studied. This is particularly true for scattering based experiments, where a significant part of reciprocal space needs to be sampled in order to be able to perform reliable Fourier transforms to real space. The situation is more favorable for direct X-ray imaging experiments, but in this case the spatial resolution is limited to a few tens of micrometers at best. Another limitation for some in-situ and time-resolved experiments is the use of complicated sample environments, like furnaces, or reaction cells, which requires high energy X-rays with sufficient penetration power. The increase in X-ray energy means a dramatic decrease in the scattering power of the sample, increasing the required exposure times.

The available photon flux at the sample has increased exponentially over the last decades; with the improvement of storage ring based synchrotron sources (see following section). However, time-resolved and in-situ experiments generally require dedicated and often complicated sample environments, which are difficult to develop for an external user of a user-facility where access to measurement time is severely limited. Real progress and groundbreaking results have been obtained at synchrotrons in the cases where an experimental station is fully dedicated to specific time-resolved or in-situ experiments. One example is the time-resolved beamline ID09 at the European Synchrotron Radiation Facility in Grenoble, France. With the increase of the number of synchrotron storage rings and experimental stations worldwide this situation is expected to improve in the future.

Another reason for the relatively low number of time-resolved and in-situ experiments as compared to the number of static and ex-situ experiments is the lack of suitable X-ray detectors. Traditionally, the development of X-ray detectors has been lacking behind the development of the storage ring X-ray sources. This is not so much because of a lack of detector developments, but rather because of the exponential increase in source brilliance. Furthermore, every facility has only one source, but many tens of different X-ray detectors. A number of detector systems have been developed specifically for time-resolved and in-situ experiments and some of these will be discussed in this chapter.

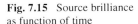

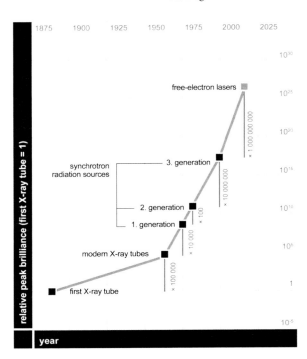

Fig. 7.15 Source brilliance as function of time

7.2.2 Storage Ring Based X-ray Sources

7.2.2.1 Source Developments

X-ray storage ring sources have developed spectacularly over the last 30 years, thanks to progress in storage ring accelerator technologies. This is best illustrated by the source brilliance as function of time given in Fig. 7.15, which shows an increase by one order of magnitude every 3 years.

Brilliance is not equivalent to flux, since it includes the sources size as well as the beam divergence, which both have decreased dramatically over time. However, it is the most relevant parameter in many experiments, since a high brilliance beam allows focusing many photons onto a small spot. This is important since either the samples themselves are small (micrometers to millimeters), or only a small volume of the sample is to be investigated. It is difficult, if not impossible, to translate the evolution of the source brilliance into an evolution of the number of photons in the detector, or detector volume. But even though a significant part of the increase in source brilliance has been offset by a parallel decrease in sample size and an evolution towards weaker scattering events, it is unambiguous that there has been a correlated increase in the flux on the detector. This increased measured intensity has made possible a corresponding decrease in exposure times, which opened the way for many time-resolved and in-situ X-ray experiments. Another important development, besides this exponential increase in source brilliance, is the increase in number of sources, and thus available experimental stations worldwide.

What follows in the Sect. 7.2.2.2 are some illustrative examples of 2-dimensional X-ray detectors used in in-situ and time-resolved experiments at synchrotron sources. A distinction has to be made between so-called "integrating" and "photon counting" detectors. Integrating detectors, integrate the total signal, produced by the X-rays as well as by the noise, during a user determined integration time. Since there is no signal processing during the integration period large fluxes can be recorded. This is not to be confused with read-out speed. For example, the read-out speed of CCDs is relatively low, whereas they can record large instantaneous fluxes. The down side is that the noise is also integrated. In photon counting detectors, the signal generated is amplified and compared to a threshold, providing a means to discriminate between noise and photons, and thus providing near noise free detectors. The down side in this case is the time it takes to amplify and process the signal, thus limiting the number of counts per second that can be recorded.

7.2.2.2 Examples of Detectors for In-situ Experiments at Storage Rings

In this chapter we will give three examples where the X-ray detector has played an essential role in enabling in-situ or time-resolved experiments. It is impossible, and by no means attempted, to be complete or to give a balanced representation, and many examples that could be included are omitted. We will concentrate on X-ray scattering experiments, but would like to stress that both imaging and spectroscopy techniques have very successfully been used in time-resolved and in-situ experiments. Furthermore, we will limit ourselves to 2-dimensional solid-state detectors.

The Cornell AP-HPAD

One of the first hybrid pixel detector systems specifically developed for fast in-situ and time-resolved experiments at synchrotron sources is the Analogue Pipeline Hybrid Pixel Array Detector (AP-HPAD), developed by the group of Sol Gruner at CHESS/Cornell [38–40]. The system was designed for fast time-resolved imaging experiments, with micro-second framing times. In order to get statistically meaningful data within micro-seconds, photon counting is not an option and an integrating detection scheme is mandatory. To optimize the efficiency of the experiments 8 consecutive images can be recorded and stored within the pixel, before reading out the detector. The schematic pixel layout of the system is shown in Fig. 7.16, and has been the basis for other systems including the Adaptive Gain Integrating Pixel Detector (AGIPD) detailed below.

Early small-scale prototypes of this system have been successfully used for high-speed radiography experiments of fuel injectors [41, 42]. In these experiments a high-pressure common-rail diesel injection system typical of that of

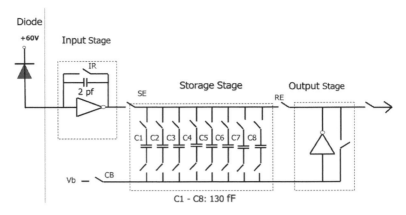

Fig. 7.16 Pixel layout of the Cornell analogue pipeline HPAD chip, [40]

a passenger car was used to study the ultra-sonic and shock wave behavior of the diesel fuel vapor immediately after exiting the orifice. The details of the three-dimensional density distribution as well as the transient behavior turned out to be considerably more complex then originally expected. The microsecond time resolution of the detector as well as the perfect repeatability and triggerability of the system under study were essential for this experiment.

The PILATUS detector

The PILATUS detector was developed by the Paul Scherer Institute for protein crystallography at the Swiss Light Source [43]. In contrast to the above described AP-HPAD, it is a photon counting detector, which has the advantage of near zero noise and excellent stability in time, as explained in the introduction. The drawback is that photon counting automatically limits the instantaneous flux that can be recorded. With a maximum count rate per pixel of the order of 10^6 counts per second, exposure times of a microsecond, used in the previous example, will yield statistically meaningless intensities. However, the system is well suited for both time-resolved and in-situ experiments. The PILATUS can be gated fast enough to isolate a single bunch of for instance the Advanced Photon Source operating in 24 bunch mode, with 153 ns separation between consecutive bunches [44]. This results in a time resolution determined by the X-ray pulse length. Similar time resolutions have been obtained before by isolating single pulses using a series of fast mechanical shutters, including fast spinning disks with slotted holes [45], which is only possible on a dedicated experimental station. The gated PILATUS detector can then be used in pump-probe experiments, where a pump, usually an optical laser, excites the sample, and a given time delay later the X-ray pulse measures the state of sample. This is then repeated many times, until enough statistics has been

Fig. 7.17 Experimental
setup in-situ rheometer
with PILATUS 300K pixel
detector

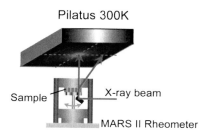

accumulated, after which the delay between pump and probe is changed, and the measurement repeated. The possibilities have been demonstrated in an experiment performed on polycrystalline organic thin films of α-perylene [44].

The PILATUS detectors have been used in a large number of in-situ experiments, including surface diffraction, powder diffraction and small angle scattering. We present here a recent experiment where small angle X-ray scattering is combined with in-situ rheology [46]. Small angle scattering gives access to nanoscopic length scales, and is well suited for millisecond and slower processes. Rheology probes the viscoelastic properties of fluid materials under steady or oscillatory shear conditions. Struth et al. used a vertical scattering geometry in order to have access to otherwise inaccessible sample orientations. The unique setup including a specially designed X-ray optical element and a modified HAAKE MARS II rheometer gives enough freedom to vary the sample-to-detector distance to the desired value (see Fig. 7.17). A PILATUS 300K system was used to record two-dimensional scattering patterns. The unique combination of in-situ rheology and a fast area detector resulted in the observation of new states of liquid crystal 8CB under nonlinear shear conditions [46]. Due to the noise free characteristics of the Pilatus module, even the weakest signals from the sample could be detected in reasonable time scales. Since beam damage is always an issue for soft condensed matter materials, fast detection times are crucial for such kind of experiments. On the other hand the time resolution in this experiment was not limited by the speed of the detector but by the flux of the source.

7.2.3 Free-Electron Laser Based X-ray Sources

7.2.3.1 Source Developments

As shown in Sect. 7.1.2.1, and Fig. 7.15, storage ring based synchrotron X-ray sources have seen an exponential increase in brilliance, revolutionizing X-ray science as well as X-ray detectors. However, this trend will not continue with storage ring based sources, as the newest sources are already very close to what is considered as the "ultimate storage ring". The next generation of sources are Free-Electron Lasers (FEL), using linear accelerators instead of circular storage rings. A thorough discussion of source and accelerator physics is far beyond the scope of

Fig. 7.18 Peak brilliance of
FEL sources as compared to
storage ring sources

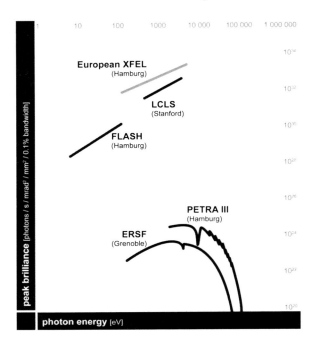

this chapter, but it is important to discuss some of the principles involved, as they
have crucial consequences for the beam characteristics and thus also for the X-ray
detectors involved.

The FEL sources are based on the Self-Amplified Spontaneous Emission
(SASE) principle, where the electrons inside a single bunch interact with their own
radiation field while traversing the insertion devices. As a result the electrons start
emitting radiation in phase, with the result that the total intensity becomes proportional to the square of the number of electrons in a bunch. This is to be compared to storage ring sources where the electrons emit incoherently, and thus the
intensity is only proportional to the number of electrons in the bunch. A number
of conditions have to be fulfilled in order for the electrons to interact with their
own radiation field, and emit coherently. Firstly, the electron bunches have to be
very compact, in order to have a sufficiently large electron density. This is hard to
achieve in storage rings, which are equilibrium sources. In single-pass linear accelerators there is no need for equilibrium, and the electron bunch can be highly compressed. As a result, not only is the generated X-ray pulse very intense, it is also
very short, typically in the femtosecond range, which is to be compared to tens of
picoseconds for storage rings. In summary, the photon pulse length is more than a
1000 times shorter than at classical 3rd generation synchrotron storage rings, the
horizontal emittance is by a factor of 100 smaller, and the vertical emittance by
a factor of 3, while the number of photons per pulse is more than a factor of 300
higher, and the natural monochromaticity a factor of 10 better, giving an increase
in peak brilliance by more than 9 orders of magnitude (Fig. 7.18). Another major

difference is the fact that the FEL beams are fully laterally coherent. It is evident that these improvements will allow for new science to be performed with the consequence that new types of X-ray detectors are required [47].

7.2.3.2 Requirements for X-ray Detectors for FEL Based Sources

A detailed description of the specifications for the X-ray detectors for Free Electron Laser sources based on scientific requirements can be found in [47]. However, it is important to point out that the field of X-ray FEL science is still in its infancy and developing rapidly, meaning that scientific requirements and detector specifications will evolve over the coming years. Nevertheless, a few specifications, resulting from the high intensity per pulse, can safely be given, and will be discussed here.

The high intensity per pulse will not only allow for, but often impose single-shot experiments. Due to the high flux of the refocused beam, samples will often not survive a single pulse because of ionization and subsequent coulomb-explosion [48]. However, since the pulse is very short, the sample will provide a full diffraction pattern before sample disintegration sets in, the so-called diffract-before-destroy principle [49, 50]. As a consequence X-ray detectors will have to record a complete diffraction image for every X-ray pulse. This means that, first of all, the X-ray detectors will have to cover a sufficiently large part of reciprocal space, with sufficiently fine pixellation. Both the size of the detector and the number of pixels required depend on the targeted scientific application [47], but generally a few million pixels of few hundred microns are required. Since a full diffraction pattern, spanning a large dynamic range, needs to be recorded per pulse (<100 fs) photon counting is excluded, leaving integrating detectors as the only option. In many experiments it is essential to distinguish between 0 and 1 photon, meaning that the detector needs to be low noise, which is particularly challenging for integrating detectors. At the same time the most intense pixels need to be able to handle more than 10^4 photons per pulse. For single-shot experiments with randomly oriented samples it is not known a priori which pixels will see low and which high intensity. Finally, the intense beam will generally prohibit the use of beam stops in front of the detector, as is customary in storage ring source experiments, and consequently the detector will need to have a central hole for the direct beam to pass through.

The requirements given above are valid for all X-ray Free-Electron Lasers, independent of the exact time structure. The European XFEL, with its super-conducting accelerator technology, presents an additional challenge, as compared other projects using so-called warm technology (LCLS in the USA and SCSS in Japan). In Fig. 7.19 the time structure of the European XFEL is shown. Bunch trains, with up to 2700 bunches separated by 220 ns are repeated 10 times per second. Since a mega-pixel detector cannot be read out in 220 ns, images will have to be stored inside the pixel for readout during the 99 ms inter-train periods. This significantly complicates the pixel design, as well as it limits the minimum pixel size obtainable.

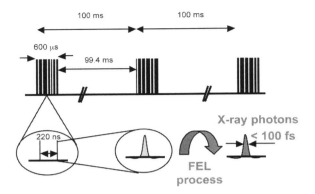

Fig. 7.19 Time structure of the European X-ray free electron laser in Hamburg

There are a number of detector development projects for X-ray FELs world-wide and a detailed technical description of these can be found in [51 and references therein]. The system most used to date, however, is the pnCCD adapted from astronomy applications [37] and as discussed also in the Sect. 7.2.3.1 on electron detectors—novel direct electron detectors.

For the European XFEL there are three projects ongoing, each one attempting to achieve the large dynamic range and the in-pixel frame storage in different technical ways. The Large-Pixel Detector (LPD) project uses three gains (high/medium/low) in parallel, and stores the images in an analogue pipeline [52]. The DEPFET Sensor with Signal Compression (DSSC) project uses a non-linear response of the sensor, and a digital memory for image storage [53]. The Adaptive Gain Integrating Pixel Detector (AGIPD) project uses an automatically adapted gain to cover the large dynamic range, and an analogue memory for image storage [54]. The AGIPD project will be presented in more detail in the Sect. 7.2.3.3.

7.2.3.3 The Adaptive Gain Integrating Pixel Detector Project

The AGIPD detector is a Hybrid Pixel detector with a silicon diode array bump-bonded to a pixellated readout chip (Application Specific Integrated Circuit, ASIC) and is being designed and built by a consortium consisting of DESY in Hamburg, the Swiss Light Source at the Paul Scherer Institute in Switzerland and the Universities of Hamburg and Bonn (Germany). Dynamic gain switching is used to provide the required large dynamic range. In this concept each pixel automatically adjusts the gain of its pre-amplifier according to the incoming signal strength, without any external intervention, and for every X-ray pulse in the bunch train. There are three possible gain settings: high, medium and low, which together cover a dynamic range from single photon sensitivity to 10^4 photons of 12 keV. As stated above, the X-ray pulses in the bunch train are separated by 220 ns (4.5 MHz), which is too short to read out the entire detector. Images are therefore stored in an analogue memory, which is randomly accessible. This last feature is important, since at most

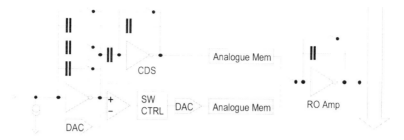

Fig. 7.20 Schematic layout of the AGIPD pixel cell

350 images can be stored, which is significantly less than the possible maximum of 2700 images available from the European XFEL source. The detector is therefore laid out for vetoing and image overwriting, such that only good images are stored. The analogue memory is then read out and digitized during the 99.4 ms train intervals. A schematic layout of the pixel cell is given in Fig. 7.20.

Detailed measurements on various prototypes are still under way, but the essential gain switching as well as the random access analogue memory have been proven to work. The current pixel size of 200 μm is a compromise between the scientific push for small pixels and the technological limitations to integrate enough functionality and storage capacity in the pixel. A single readout ASIC, the fundamental building block, will have 64 × 64 pixels. The ultimate 1k × 1k pixel detector will be constructed using monolithic multi-chip modules, with 2 × 8 readout chips (128 × 512 pixels), as indicated in Fig. 7.21, and is foreseen to be ready by 2014.

Although AGIPD, is specifically developed for the European XFEL, with its unique time structure (see Fig. 7.19), it also offers great opportunities for time-resolved and in-situ X-ray experiments at storage ring sources. Since it is an integrating detector, it does not suffer from count rate limitations, which often limits the time resolution when using photon counting detectors. At the same time, up to 350 images can be recorded in very fast succession, down to 220 ns or 4.5 MHz. This could be used in, for instance, pump-probe experiments, where a trigger initiates a reversible or repeatable reaction and the evolution of sample is then followed in 220 ns intervals. This can then be repeated 10 times per second. Since the operation of the detector is fully triggered by external signals, there is a considerable freedom in the sampling rate. For instance one could follow the sample shortly after the pump has initiated a change at 220 ns time intervals, and gradually increase this spacing. There are, however, some technological limits to the degrees of freedom. Since an analogue memory is used to store the signal inside each pixel, one has to read out the images within a given time, otherwise the stored signal will slowly fade away due to the so called signal droop. The full extend of the limitations of the degrees of freedom will only become clear once the final readout ASIC is ready and fully tested.

It should be pointed out that, although developing a system like AGIPD is a multi year and multi million Euro project, deriving variations, optimized for storage ring applications, is considerably faster and cheaper.

Fig. 7.21 Layout of the final 1k × 1k detector, featuring a central hole for the primary beam

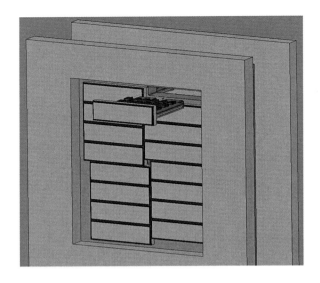

7.2.4 Outlook

X-ray detectors have traditionally been the weakest part in many synchrotron-based experiments, especially in time-resolved and in-situ experiments. This is not due to a lack of progress in detector technology, but rather due to the phenomenal increase in source brilliance over the last decades. Since this brilliance increase is leveling of for storage ring based X-ray sources, an increased emphasis is put on detector development. Although it is impossible to give an accurate prediction of the future, a few directions and expected developments should be mentioned.

For synchrotron experiments there is a clear shift away from using or adapting detectors or components developed for other scientific or non-scientific applications, towards custom made detector systems. A good example is the PILATUS based system developed specifically for protein crystallography experiments [43]. This trend is continuing with, for example, the AGIPD development mentioned above. The hybrid pixel array detector technology with Application Specific Integrated Circuits (ASICs), will show a trend towards using more intelligence inside the pixel, like the automatically adapting gain or communicating pixels [55, 56]. Using new developments in industry, like the 3-dimensional integration of CMOS technology [57] one can dramatically increase the functionally per unit area, or alternatively decrease the pixel size, while keeping full functionally. This 3D-integration will also open the way for building 4-side buttable detector modules, which can be tiled together into large detector systems, with negligible dead areas. In parallel to the developments in the readout electronics, we are seeing rapid developments in the sensor technology, like 3D and edgeless silicon [58] or high-Z semiconductors [59] for high photon energies. With avalanche diode arrays or silicon PMTs one will be able to reach time resolutions down to the nano-second level [60], while low noise, and thus high-energy resolution can be obtained with DEPFET based pixel sensors [53]. In the coming years we will see various combinations of the above possibilities. It might not be impossible that we will be able to determine for every recorded photon its energy with near Fano limited resolution, and assign it to the electron bunch that generated it, thus reaching machine limited energy, as well as time-resolution in the long term.

References

1. C. Ponchut, J. Synchrotron Radiat. **13**, 195 (2006)
2. J.D. Dainty, R. Shaw, *Image Science* (Academic Press, London, 1974)
3. K.H. Herrmann, D. Krahl, *Advances in Optical and Electron Microscopy*, vol. 9 (Academic Press, London, 1984), p. 1
4. K. Ishizuka, Ultramicroscopy **52**, 7 (1993)
5. J.M. Zuo, Ultramicroscopy **66**, 21 (1996)
6. A.L. Weickenmeier, W. Nüchter, J. Mayer, Optik **99**, 147 (1995)
7. O.L. Krivanek, P.E. Mooney, Ultramicroscopy **49**, 95 (1993)
8. W.J. de Ruijter, J.K. Weiss, Rev. Sci. Instrum. **63**, 4314 (1992)
9. K. Downing, D.A. Grano, Ultramicroscopy **7**, 381 (1982)
10. P.J.W. Noble, IEEE Trans. Electr. Dev. **ED15**, 202 (1968)
11. S.G. Chamberlain, IEEE J. Sol. Stat. Circ. **SC4**, 333 (1969)
12. P.K. Weimer, W.S. Pike, G. Sadasiv, F.V. Shallcross, L. Meray-Horvath, IEEE Spectr. **6**, 52 (1969)
13. E.R. Fossum, Proc. SPIE **1900**, 2 (1993)
14. B. Dierickx, G. Meyants, D. Scheffer, in *Proceedings of IEEE CCD & AIS Workshop* (1997), p. P1
15. G. Deptuch, Nucl. Instrum. Meth. Phys. Res. A, **543**, 537 (2005)
16. G. Mettivier, Nucl. Instrum. Meth. Phys. Res. A, **516**, 554 (2004)
17. S.R. Amendolia, et al., Nucl. Instrum. Meth. Phys. Res. A, **466**, 74 (2001)
18. R.H. Richter et al., Nucl. Instrum. Meth. Phys. Res. A, **511**, 250 (2003)
19. M. Battaglia et al., Nucl. Instrum Meth. Phys. Res. A, **608**, 363 (2009)
20. G. McMullan et al., Ultramicroscopy **107**, 401 (2007)
21. P. Bartl et al., Nucl. Instrum. Meth. Phys. Res. A, **591**, 314 (2007)
22. R. Turchetta et al., Nucl. Instrum. Meth. Phys. Res. A, **458**, 677 (2001)
23. J. Matheson et al., Nucl. Instrum. Meth. Phys. Res. A, **608**, 199 (2009)
24. A. Blue et al., Nucl. Inst. Meth. Phys. Res. A, **581**, 287 (2007)
25. H.S. Matis et al., IEEE Trans. Nucl. Sci. **50**, 1020 (2003)
26. N.H. Xuong et al., Proc. SPIE-IS&T Elect. Imag. **5301**, 242 (2004)
27. G. Varner et al., Nucl. Instrum. Meth. Phys. Res. A, **541**, 166 (2005)
28. G. McMullan, S. Chen, R. Henderson, A.R. Faruqi, Ultramicroscopy **109**, 1126 (2009)
29. M. Deveaux et al., Nucl. Inst. Meth. Phys. Res. A, **583**, 134 (2007)
30. J. Bogaerts, B. Diericks, G. Meynants, D. Uwaerts, IEEE Trans. Electr. Dev. **50**, 1 (2003)
31. M. Deveaux et al., Nucl. Inst. Meth. Phys. Res. A, **552**, 118 (2005)
32. E.G. Villani, M. Turchetta, M. Tyndel, Nucl. Phys. B **125**, 184 (2003)
33. L. Strüder et al., Astron. Astrophys. **365**, L18 (2001)
34. W. Leitenberger et al., J. Synchrotron Radiat. **15**, 449 (2008)
35. R. Hartmann et al., Nucl. Instrum. Meth. Phys. Res. A, **568**, 188 (2006)
36. N. Meidinger et al., IEEE Trans. Nucl. Sci. **45**, 2849 (1998)
37. L. Strüder et al., Nucl. Instrum. Meth. Phys. Res. A, **614**(3), 483 (2010)
38. S.L. Barna et al., IEEE Trans. Nucl. Sci. **44**, 950 (1997)

39. E.F. Eikenberry et al., J. Synchrotron Radiat. **5**, 252 (1998)
40. G. Rossi et al., J. Synchrotron Radiat. **6**, 1096 (1999)
41. A.G. MacPhee et al., Science **295**, 1261 (2002)
42. W. Cai et al., Appl. Phys. Lett. **83**, 1671 (2003)
43. P. Kraft et al., IEEE Trans. Nucl. Sci. **56**, 758 (2009)
44. T. Ejdru et al., J. Synchrotron Radiat. **16**, 387 (2009)
45. M. Wulff et al., Faraday Discuss. **122**, 13 (2002)
46. B. Struth et al., Langmuir **27**, 2880 (2011)
47. H. Graafsma, JINST **4**, P12011 (2009)
48. R. Neutze, R. Wouts, D. van der Spoel, E. Weckert, J. Hajdu, Nature **406**, 752 (2000)
49. H. Chapman et al., Nature **470**, 73–77 (2011)
50. M.M. Seibert et al., Nature **470**, 78–81 (2011)
51. H. Graafsma, *Semiconductor Radiation Detection Systems* (CRC-Press, Boca Raton, 2010), ISBN: 9781439803851
52. A. Blue, M. French, P. Seller, V. O'Shea, Nucl. Instrum. Meth. Phys. Res. A, *607*, 55–56 (2009)
53. M. Porro et al., in *IEEE Nuclear Science Symposium Conference Record* (2008), p. 1578
54. X. Shi et al., Nucl. Instrum. Meth. Phys. Res. A, **624**, 387 (2010)
55. R. Ballabriga, M. Campbell, E. Heijne, X. Llopart, L. Tlustos, IEEE Trans. Nucl. Sci. **NS-54**, 1824 (2007)
56. D. Pennicard, R. Ballabriga, X. Llopart, M. Campbell, H. Graafsma, Nucl. Instrum. Meth. Phys. Res. A, **636**, 74 (2011)
57. International Technology Roadmap for Semiconductors (ITRS) report 2009, and 2010 update interconnection section; www.itrs.net
58. S.I. Parker, C.J. Kenney, J. Segal, Nucl. Instrum. Meth. Phys. Res. A, **395**, 329 (1997)
59. D. Greiffenberg, A. Fauler, A. Zwerger, M. Fiederle, JINST **6**, C01058 (2011)
60. C. Thil et al., Nucl. Instrum. Meth. Phys. Res. A, **628**, 461 (2011)

Index

A. Ziegler et al. (eds.), *In-situ Materials Characterization*,
Springer Series in Materials Science 193, DOI: 10.1007/978-3-642-45152-2,
© Springer-Verlag Berlin Heidelberg 2014

Printed by Publishers' Graphics LLC
MLSI140421.15.20.102